Copyright © AceOrganicChem.com 2013, all right reserved

ISBN 978-1484909652

Without limiting the right under copyright reserved above, no part of this publication may be reproduced, stored in or introduced into a retrieval system, or transmitted in any form or by any means (electronic, mechanical, photocopying, or otherwise) without express written consent of AceOrganicChem.com or its designee and the publisher of this book.

A quick word from our sponsor:

On the web: AceOrganicChem.com is your one-stop on the interwebs for everything you need to help you get a great grade in organic chemistry. With videos, e-books, flashcards, a test bank, a tutorial bank, a best of the web, MCAT help and a little bit of chemistry humor, AceOrganicChem.com has it all. Much of the site is free and is helpful for students of all levels and abilities.

On Facebook: Find us by searching "organic chemistry help" or "AceOrganicChem." Like us and you will get updates when free stuff comes available, see interesting articles on new discoveries in chemistry, and catch a joke every once in a while. We will not inundate you with crap, but you will get the word on free help from us before anyone else.

On Google Plus: Same deal as Facebook. We will not inundate you with crap, but you will get the word on free help from us before anyone else.

AceOrganicChem.com
You can't fight organic chemistry, but you can ace it.

Dedicated to my lovely wife and our wonderful Katiebug, Moochie Monster, and Teddy Bear.

Ace Organic Chemistry Mechanisms with EASE

Table of Contents

Pre-Game Warm-ups: 7

    Introduction: 8
    The Language 9
    Resonance 10
    Arrows Will Point the Way 12
    Functional Groups 13
    A Brief Preview of the EASE Method: 16

1st Quarter: 18

    STEP 1--"E" is for Electrophile: 19
        Electrophiles: 19
        Leaving groups create electrophilic sites: 20
        Nucleophiles: 23
    Step 2--"A" is for Acid: 32
        Acids: 32
        Acids from Unlikely Places: 34
        Lewis Acids: 35
        Bases: 38
    Step 3 --"S" is for Sterics:. 44
    Step 4 --"E" is for Electron Flow: 50

2nd Quarter: 57

    Need to Know Mechanism #1: SN1 58
    Need to Know Mechanism #2: SN2 60
    Need to Know Mechanism #3: E1 62
    Need to Know Mechanism #4: E2 65
    Need to Know Mechanism #5: Electrophilic Aromatic Substitution .. 67
    Intermediate Mechanism #1: Nucleophilic Addition to a Carbonyl. 69
    Intermediate Mechanism #2: Alkene Addition 71

- Intermediate #3: Dieckmann Reaction 73
- 7 More Mechanisms All-Stars Will Know: 75

3rd Quarter: 76

- Easy Problem: What is the Mechanism? 77
- Intermediate Problem: What is the Mechanism? 78
- Hard Problem: What is the Mechanism? 80
- Easy Problem: What is the Product? 87
- Intermediate Problem: What is the Product? 90
- Real Chemistry You Have Never Seen Before: 93

4th Quarter: 97

- Your Organic Chemistry Toolbox: 98
- Let's Go Retro: Retrosynthesis 100
- Spiderwebs: 103

Overtime: 105

- Where the method will be less effective: 106
  - Oxidation and Reduction: 106
  - Radical Reactions: 108
  - Organometallic Reagent Formation: 109
- Appendix 1: Answers To Supplemental Exercises 110

Pre-Game Warm-ups:

The Basics

Introduction:

Let me state the obvious first: This is a book designed to help you learn how to do organic chemistry mechanisms and synthesis problems.

Now let me state the less obvious: This is **not** supposed to be a comprehensive organic chemistry textbook. It is a supplement to your lecture and practice questions, but is not meant to replace them. In fact, your textbook and lecture notes are an essential prerequisite to understanding this book and using it properly.

We have found that using sports metaphors to teach organic chemistry helps all students, even ones who don't watch or like sports. Therefore, organic chemistry is much like going to a football game:

1) You will need to understand the language before you can fully understand what is going on in the game. Believe it or not, organic chemistry is a foreign language to everyone when they first start. Initially, as your professor teaches a new topic, you will need to translate it in your head to properly understand what is going on. However, as you progress and become more fluent, you will not need to translate everything, you will just understand it.
2) You will need to know all of the players. Just like on the soccer pitch, if you don't know what each reactant is supposed to be doing, you won't know what is going to happen in the reaction.
3) You will need to know which molecules are going to be on attack (the offense) in which molecules are going to be attacked (the defense). For our analogy, molecules that attack will have electrons and molecules that get attacked are electron deficient.
4) All of the players sort of smell bad. Just like the French National Men's Soccer Team, organic chemicals are notorious for their odors.

This text takes a methodical approach to teaching mechanisms and synthesis problems. First, we will review the new language (called organic chemistry) that you must learn, although you will have seen most of this through your textbook and lecture class. We will then teach you about all of the players, and finally show you how it all fits together and why reactions occur.

By the end of the book it is our goal to not only teach you how to write organic chemistry mechanisms, but also teach you how to think through problems you may have never seen before. You will be able to do this because the rules of organic chemistry will never change; you just need to know how each player is going to react in a given situation. Knowing this will allow you to think through any synthesis/mechanism problem which might get thrown at you.

Disclaimer: While the EASE method is a great resource and forces you to think about the electronics of a reaction, it is not always the best way to proceed. We will show you reactions in the Overtime section where it is just plain easier to know what the reagent does, and disregard how/why it works. One glaring place where this is true is in oxidation and reduction chemistry, as these mechanisms can be very complicated and are most likely not something an undergraduate professor would want you to know for an exam. Therefore, we suggest in order to maximize your study time, it is best to just know what the product of these reactions will be, and not worry about that mechanism.

Overall though, we think you'll find this method will be incredibly helpful and will assist you in figuring out the mechanisms of reactions that you may not have seen before.

So without further hesitation, let's get to that organic chemistry football game.

The Language

Learning organic chemistry is like trying to work in a foreign country; if you don't know the language, it is going to be very difficult to learn how to do your job. Imagine that you have just been transported to the mythical country of "ochemia," a small island nation in the south Pacific, where your job is to write chemical reactions.

Frequently, in a chemistry lecture, professors start tossing out strange organic chemistry terms far too rapidly. Because students aren't fluent in "ochemia" yet, they need to translate each word in their head to understand what the instructor has just said. By the time this mental translation is done, the student has just missed the next sentence and has lost half of the lecture. Our goal is to get as fluent as we can in the language of chemistry as quickly as we can. Here are _some_ terms it will be helpful to memorize so that you don't have to do a mental translation when you hear them:

Meth = 1

Eth = 2

Prop = 3

But = 4

Pent = 5

Hex = 6

Hept = 7

Oct = 8

Non = 9

Dec = 10

Nucleophile = has electrons, has a negative or partial negative charge

Electrophile = wants electrons, has a positive or partial positive charge

Halogen = F, Cl, Br, I

Proton = H⁺ or a hydrogen attached to a molecule that could come off

Aprotic solvents = do not contain OH or NH bonds

Protic solvents = contain OH or NH bonds

Bronsted Acid = proton donor

Bronsted Base = proton acceptor

Lewis Acid = electron acceptor

Lewis Base = electron donor

Carbonyl group = [C=O]

Cis = same side of a double bond or ring

Trans = opposite sides of a double bond or ring

Reactant = A starting material that will undergo a chemical reaction

Resonance

Resonance is one of those issues that you will have to deal with for both semester I & II of organic chemistry. You will see it as an exam topic over and over again. It is much better to have a solid understanding of it now, rather than have to worry about it later.

Moreover, resonance will help show which sites might be nucleophilic and which might be electrophilic. This is huge for helping to figure out the mechanisms and products of reactions which we may have never seen before.

The basic goal of resonance structures is to show that molecules can shift electrons and charges onto different atoms on the same molecule. This makes the molecule generally more stable because the charge is now delocalized and not "forced" onto an atom that does not want it.

Below are some handy rules of resonance. If you learn these and think about them when tackling different resonance problems, you will be able to handle whatever is thrown at you.

1) Know each atom's "natural state." You need to recognize what each atom generally looks like, in an uncharged state. Below is a brief summary of some important atoms in organic chemistry and what their uncharged state is:

In an UNCHARGED state:

- C has four bonds (and no lone pairs)

- N has three bonds (and one lone pair)
- Halogens have one bond (and three lone pairs)
- O has two bonds (and two lone pairs)
- H has one bond (and no lone pairs)

Three more rules:

- C and N are central atoms, meaning that they will usually be in the middle of your molecule, usually not as a terminal atom (IF they are neutral or single-bonded)
- Hydrogen (H) and halogens (X) are almost always terminal atoms, meaning that they will only have one bond and be at the ends of molecules.
- With the exception of H, atoms in group I & group II are only counterions (+1 or +2 charge and not involved in resonance).

These rules will help you to construct the Lewis Dot structure on which you will base your resonance structures. Remember that halogens and hydrogens are always terminal, meaning that are at the end of the molecule and only have one bond, and therefore, they will not participate in resonance.

2)   Atom positions will not change. Once you have determined that an atom is bonded to another atom, that order will not change in a resonance structure. If they do change, it is no longer a resonance structure, but is now a constitutional isomer or a tautomer.

3)   Check the structure you have created to make sure that it follows the octet rule. This will become much easier once you have a better handle on the "natural state" of atoms.

4)   When two or more resonance structures can be drawn, the one with the fewest total charges is the most stable. In the example below, A is more stable than B.

5) When two or more resonance structures can be drawn, the more stable has the negative charge on the more electronegative atom. In the example below, A is more stable than B.

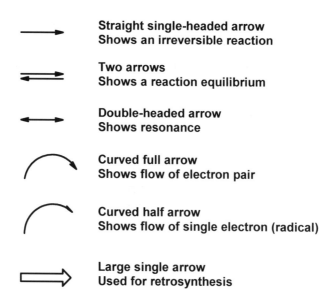

Most importantly, in the end, each resonance structure should have the same overall charge and total number of electrons (bonds + lone pairs) as when you started.

At various times in the book we will expand on this theme and show you how resonance contributes to the electrophilic/nucleophilicity of a structure. This will be a major clue for why some molecules behave as they do.

Arrows Will Point the Way

While it may seem like a minor point, different arrows in organic chemistry mean different things:

→ **Straight single-headed arrow**
Shows an irreversible reaction

⇌ **Two arrows**
Shows a reaction equilibrium

↔ **Double-headed arrow**
Shows resonance

⌒ **Curved full arrow**
Shows flow of electron pair

⌒ **Curved half arrow**
Shows flow of single electron (radical)

⇨ **Large single arrow**
Used for retrosynthesis

The most important thing to understand with these arrows is the electron flow with curved arrows. This will be an integral part of any mechanism which you will draw. As a strict rule, electron flow is *always* drawn showing where the electrons are *moving to*. This means that electrons will flow from negative to positive, from nucleophile to electrophile, or from base to acid.

Functional Groups

Let's not beat around the bush on this one: functional groups are why we can do organic chemistry. Functional groups inherently contain polar bonds and are the basis of why molecules can react with each other. Without functional groups, everything would be straight chain alkanes and other boring hydrocarbons. So it's important to learn about functional groups, and how they will interact with nucleophiles and electrophiles.

Hydrocarbons: these are simply composed of carbon and hydrogen. This group is composed of alkanes, cycloalkanes, alkenes, and alkynes. Don't forget about conjugated alkenes too, as they are important in many organic processes such as the Diels-Alder reaction. While alkanes and cycloalkanes are not particularly reactive, alkenes and alkynes definitely are.

Carbonyls: a "carbon double bond oxygen" is a carbonyl. It is one of the more important electrophiles you will see in this course. While there are different variations which can make the carbonyl more or less reactive, the basic functional group is still the same. The important point here is to know which types of carbonyls are more electrophilic and which ones are less. Generally speaking, if there is an electron withdrawing group attached to the carbonyl carbon, that carbonyl will be more electrophilic and more reactive.

Alkyl Halides: alkanes which are connected to a halogen atom (F, Cl, Br & I) are good electrophiles. These can participate in nucleophilic substitution and elimination reactions. Their reactivity depends on the type of alkyl halide (F, Cl, Br & I), its substitution (primary, secondary, tertiary) and the desired reaction (SN1, SN2, E1, E2, or others).

Alcohols, Amines, and Thiols: these are generally very good nucleophiles, as the heteroatoms have lone pairs which will attack an electrophile.

Ethers: do not undergo many organic reactions themselves, but sometimes can be the product of a reaction. Some chemists refer to ethers as "dead molecules" because of their low reactivity.

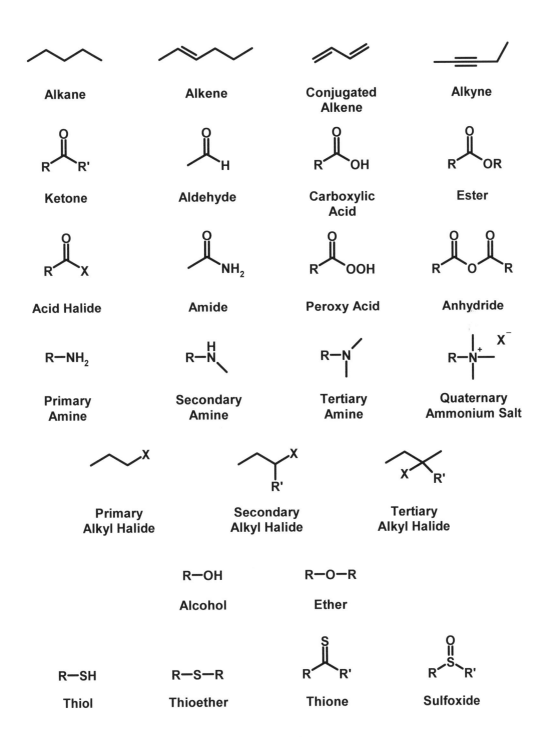

A Brief Preview of the EASE Method:

Step 1: E is for electrophile. Determine which atom(s) could be the electrophile and which could be the nucleophile and label them.

Step 2: A is for acid. Determine if an acid or base is present. If there is a strong acid or base present, move a proton before going on with the EASE method.

Step 3: S is for sterics. Determine if there are any steric impedances in your reaction. If there are, pause and reassess.

Step 4: E is for electron flow. Once you have completed steps 1-3, move the electrons in the reaction from the nucleophile to the electrophile.

The EASE method is a methodical and stepwise approach to determining the mechanism and/or products of an organic chemistry reaction. If used properly, there is not a reaction that can't be reasoned out using this method. (Of course sometimes it's easier just to memorize the outcome, but we'll discuss that in the last section). Let's briefly look at a sample mechanism using the EASE method:

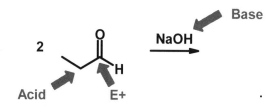

E - Step 1: E+ is obvious, unsure about Nu-
A - Step 2: NaOH is a base, alpha H is acid, move them

In the reaction above, we take two equivalents of an aldehyde and react it with sodium hydroxide, a strong base. This reaction is commonly referred to as the aldol reaction. In step 1, we need to identify the electrophile and the nucleophile. It should be obvious that the carbonyl is the electrophile, but it is unclear which is the nucleophile. Since we are unsure about this, let's skip to step 2 to see if that helps us out. In step 2, we will identify the acid and the base and move the proton, if necessary. We can quickly determine the sodium hydroxide is the base, and label the proton alpha to the carbonyl as the acid. We now abstract that proton with the base and create an enolate ion.

E - Step 1: E+ is still obvious, Nu- is deprotonated carbon
A - Step 2: Base has been used up
S - Step 3: There are no steric issues
E - Step 4: Electron flow is intermolecular

This now allows us to clearly identify our nucleophile in the problem, so we start the process over with step 1 completed. There is no more base remaining, so step 2 is null. Step 3 is to identify any steric issues, of which there are none. Step 4 is to move the electrons from nucleophile to electrophile. A clue to the electron flow can be found in the fact that the nucleophile and electrophile are on the same molecule. This means electron flow could be inter- or intra-molecular. Because the two are on adjacent carbons and cannot form a new bond, the reaction must proceed in an intermolecular fashion. Hence, once the electrons are moved from the nucleophile to the electrophile, we can see the product formed and have completed the reaction successfully.

The power of the EASE method lies in the fact that it can be used in many, many different circumstances. It is especially effective when you don't know what the products formed should be. This stepwise method allows you to methodically break down how the reaction proceeded, and allows you to determine the products of reactions that you may have never seen before.

In the coming sections, we will go more in-depth into each step of the method and teach you all the tricks to look for to figure out different organic chemistry reactions.

1ˢᵗ Quarter:

The Four Steps of the EASE Method

STEP 1--"E" is for Electrophile: Identify your electrophiles and nucleophiles.

Electrophiles:

Just like in football, it is easy to say that one of the players is the most important one in the game. While many (nerdy) organic chemists could have a robust debate over a pint as to which compound class is most valuable in a reaction, we are going to treat them all as important. In its most basic form, they are all essential in some way or another to the reaction's success. Hence, we are going to start with acids and discuss all of the compound classes individually.

That being said, electrophiles might be one of the two most important reactant classes in organic chemistry. As we discussed above, organic chemistry reactions are all about the flow of electrons, and electrophiles are the ones who want those electrons. When you think of the word "electrophile" you should think of the Greek word "philos" which means "to love." Therefore, an electrophilic species is one that "loves electrons." Easy enough, right? Since opposites attract, and the electrophile loves electrons, then it must be that the electrophile is partially or fully positively charged. Most often, you will see this abbreviated as "E^+."

So the question now becomes: what makes an atom a good electrophile and how do we spot it? Since we know that electrophiles want to electrons, the first clue that something is electrophilic is it has a positive charge. The second clue is if we can place a positive charge somewhere on the atom via resonance and that it has an empty orbital (positive charge or metal with an empty orbital) or can get an empty orbital by kicking off a leaving group. Below are some common classes of electrophiles you will frequently see in your course:

A) R-C(=O)-R ⇌ R-C$^+$(-O$^-$)-R

B) Cl—Cl ⇌ $^{\delta+}$Cl—Cl$^{\delta-}$

C) H$_3$C—Cl ⇌ $^{\delta+}$H$_3$C—Cl$^{\delta-}$

D) HCl ⟶ Cl$^-$ + H$^+$

E) NO$_3$ $\xrightarrow{H_2SO_4}$ O=N$^+$=O

A carbonyl is shown in example A. We know that the carbon of the carbonyl is electrophilic because we can place a positive charge on it via resonance. This means that a nucleophile will attack the carbonyl at this carbon atom. Diatomic chlorine is shown in example B. Diatomic halogen molecules are electrophilic because the bond between the halogen atoms as polarizable, meaning that the electron cloud can partially favor either atom at any time, making one of the atoms more electrophilic than the other. In example C, we see that alkyl halides are also electrophilic because of a polarizable bond between the carbon and the chlorine atoms. Unlike example B, example C has a permanent dipole. Example D is an example of a strong acid completely disassociating, which gives off a proton as the electrophilic species. Finally in example E, we see that you can create an electrophile from a non-electrophilic molecule. Here we have reacted nitric acid with sulfuric acid to form the nitronium ion, which is highly electrophilic.

Now that we've seen some of the more elementary electrophiles, we can move on to some more complex ones. Below are four more examples of electrophiles you will see in the future:

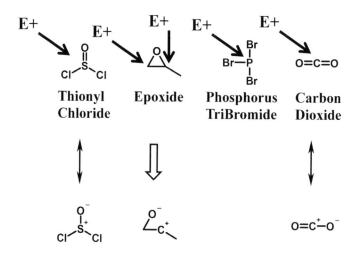

Notice that each electrophile has a positive spot that can be identified either by resonance or by examining the structure closely.

Leaving groups create electrophilic sites:

Have you ever met that person seems to always be hanging around the team, but when practice gets tough, they're the first ones to hit the showers and leave? In organic chemistry, that is a leaving group. They don't even need a good excuse to depart, they just do sometimes. So what is the one overriding similarity between that guy and a good leaving group? Simple, they both enjoy being negative.

Leaving groups are atoms and molecular fragments that can easily support a negative charge because of resonance, electronegativity, or polarizability. Below is a table of the most common leaving groups and their "nicknames.

Formal name	Structure	Nickname
Iodide	I⁻	None
Chloride	Cl⁻	None
Bromide	Br⁻	None
Methanesulfonate	H₃C—S(=O)(=O)—O⁻	"Mesylate" or "-OMs"
Toluenesulfonate	H₃C—C₆H₄—S(=O)(=O)—O⁻	"Tosylate" or "-OTs"
Trifluoromethanesulfonate	F₃C—S(=O)(=O)—O⁻	"Triflate" or "-OTf"

Halides are generally known to be good leaving groups, but lesser known are the sulfonate esters. Using resonance, these special esters can distribute the negative charge over a number of atoms, thus stabilizing it and making it labile. In fact, the sulfonate esters are between 10^3 and 10^8 times more labile leaving groups than the halides.

Also, another moiety which can be a good leaving group is the positively charged portion of a molecule. The classic example of this is when a hydroxyl (poor leaving group), becomes positively charged (aka water, a very good leaving group). See below for some examples where leaving groups are used in substitution or elimination reactions.

Carbonyls are an interesting group of electrophiles. They can be broken into three categories: "supercharged," "normal" and "boring." This designation comes from what type of functional group is connected to the carbonyl carbon. As you can see below, electron withdrawing groups (EWG) increase the electrophilicity by withdrawing electron density from the carbonyl center. This is in contrast to electron donating groups (EDG), which decrease the electrophilicity by donating electron density to the carbonyl center.

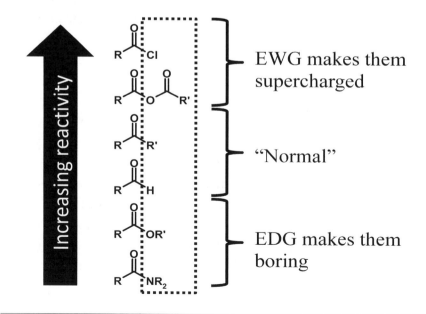

Take home points on electrophiles:

1) They want electrons, meaning they are electron deficient.
2) They are attacked by nucleophiles.
3) They are positively charged, polar and/or polarizable.
4) They become better electrophiles in the presence of Lewis acids.

Nucleophiles:

Nucleophiles are important throughout organic chemistry, but will be especially important when trying to determine the products of elimination and substitution reactions. As the name states, a nucleophile is "nucleus loving," meaning it wants to react with a positive charge. This is the core of organic chemistry: nucleophiles will attack electrophiles.

Nucleophiles are obviously very similar to bases, as there are many compounds which can be both a nucleophile and a base. Just like in sports, we want to be sure to use the right player in the right situation. Therefore, to be effective, we have to know which compounds are better bases, which ones are better nucleophiles, and which can play both positions.

There are generally three trends to remember when discussing how nucleophilic a reactant is:

1) <u>Size</u> - Generally, the more linear and/or smaller the nucleophile, the *more* nucleophilic it will be. This is because it can react at more sites and is not sterically hindered if it is smaller or linear.
2) <u>Electronegativity</u>- The more electronegative an atom is, the *less* nucleophilic it will be. This is because more electronegative atoms want to hold electron density closer, and therefore will be less likely to let that electron density participate in a reaction. We see this in calculations and experiments that show nucleophilicity decreases as you get closer to fluorine on the periodic table (C > N > O > F).
3) <u>Polarizability</u>- The more polarizable an atom is, the *more* nucleophilic it will be. Polarizability is defined as the ability to distort the electron cloud of an atom, which allows it to interact with a reaction site more easily. Generally, polarizability increases as you travel down a column of the periodic table (I > Br > Cl > F).

Below is a table of relative nucleophilic strength. This is relative because nucleophilic strength is also dependant on other factors in the reaction, such as solvent.

VERY Good nucleophiles	HS^-, I^-, RS^-
Good nucleophiles	Br^-, HO^-, RO^-, CN^-, N_3^-
Fair nucleophiles	NH_3, Cl^-, F^-, RCO_2^-
Weak nucleophiles	H_2O, ROH
VERY weak nucleophiles	RCO_2H

As shown above, as a general rule, the anion of a reactant will be a better nucleophile than the neutral form. (i.e. RCO_2^- is a better nucleophile than RCO_2H)

One very important class of nucleophiles you will see is the organometallics, shown in the figure on the next page. These are highly nucleophilic species, and will be emphasized by professors in any organic chemistry course. The key to understanding organometallics is to remember it is the carbon attached to the metal that is nucleophilic, not the metal atom itself.

We should also address aromatic rings here. Many students forget that in electrophilic aromatic substitution (EAS) it is the aromatic ring that attacks the electrophile.

Aromatic Rings (EAS)

Alkenes

Carbanions

Organometallics

The final set of nucleophilic compounds we will discuss is the pi- bond containing compounds, namely alkenes and alkynes. Some professors refer to the double bond of an alkene as a "crunchy double bond." Generally, in synthetic organic chemistry, double bonds will act as a nucleophile and grab electrophilic species, leaving a carbocation on the carbon backbone. The most obvious example of this is electrophilic addition to an alkene.

Alkyne carbanions are also highly nucleophilic. These are very useful in carbon-carbon bond formation. However, care must be taken, as these can also be strong and unhindered bases. In addition to the great nucleophilicity, these molecules also provide a very versatile functional group for later in a synthetic pathway. We will discuss more about the alkyne's acidic terminal proton in Step 2, but be aware these are great nucleophiles.

"BMOC" Nucleophile

Finally, a carbon next to a strong electron withdrawing group can also be deprotonated to create a strong nucleophile. Example of these are shown on the next page. Of note, the acetonitrile anion, which is versatile but very rare, can be used as a nucleophile when you want to stick a –CH2CN moiety on your molecule.

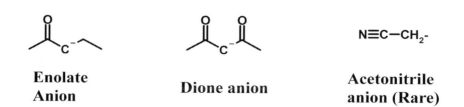

Enolate Anion   **Dione anion**   **Acetonitrile anion (Rare)**

Take home points on nucleophiles:

1) Nucleophiles attack electrophiles.
2) Nucleophiles can be basic and bases can be nucleophilic.
3) Several variables, including size, electronegativity, polarizability, and solvent will affect nucleophilicity.
4) More often than not, an alkene will be a nucleophile.

Starting Step 1:

So we have decided that step 1 is to identify the nucleophile or electrophile in the reaction. To do this, we suggest you place a large "E+" or "Nu-" near the electrophile and nucleophile, respectively. Some examples of this are below:

1) cyclopentanone (E+ at C=O) + CH₃MgBr (Nu-) ⟶

2) propyl chloride (E+ at C-Cl) + NaN₃ (Nu-) ⟶

3) methylenecyclopentane (Nu- at =CH₂) + Br₂ (E+) ⟶

4) isopropylbenzene (Nu- on ring positions) + Cl₂ (E+), AlCl₃ ⟶

In reaction 1, we have labeled the carbonyl carbon as the electrophile and the carbon attached to the Grignard as the nucleophile. In reaction 2, the alkyl halide carbon is the electrophile and the azide anion is the nucleophile. In reaction 3, the alkene is the nucleophile, with the terminal carbon being slightly more nucleophilic, due to the fact that the interior carbon is more capable of holding a positive charge. Because bromine is so electronegative, one of the bromine atoms will serve as the electrophile and be attacked by the alkene. In reaction 4, there are three potential nucleophilic sites on the aryl ring. (When the reaction finally proceeds, the most nucleophilic site will be determined by steric hindrance.)

Labeling the proper nucleophile/electrophile is essential in order to correctly determine the electron flow. We know electrons will flow from the nucleophile to the electrophile. However we won't move the electrons until step 4 of the EASE method.

One last note on nucleophiles and electrophiles: One of the strongest electrophiles is a carbocation. There are two steps that must be taken if you believe that you will be forming a carbocation in your reaction. The first is that you must determine if you have the right conditions to form the cation, which would be a tertiary (or special case) substrate and a polar protic solvent. The second step is to check to see if your cation can rearrange to a more stable cation.

<u>HERE IS A HINT</u>: Most professors who give you a carbocation problem put a rearrangement somewhere in the problem because they love that trick as they have zero creativity and have been asking that question since the dawn of time.

If you have a carbocation which is capable of rearranging (and most likely you will), you need to rearrange that cation as soon as it is formed. This may be during step 1 (call it "step 1b") or it may be during step 2. Either way, be aware of the rearrangement.

Below are some practice problems. Please label the nucleophile and the electrophile in these reactions.

1) cyclopentanol −OH + $H_3C-C(=O)-Cl$ ⟶

2) CH$_3$CH$_2$CH$_2$−Br + CH$_3$SH ⟶

3) phenyl−MgCl + O=C=O ⟶

4) 3-methylphenol (H$_3$C−, −OH on benzene) + acetic anhydride $\xrightarrow{H_2SO_4}$

Answers to practice problems:

1) cyclopentyl-OH (Nu-) + H₃C-C(=O)-Cl (E+) ⟶

2) CH₃CH₂CH₂-Br (E+ on carbon) + CH₃SH (Nu-) ⟶

3) Ph-MgCl (Nu- on C) + O=C=O (E+) ⟶

4) 3-methylphenol: H₃C–C₆H₄–OH (Nu-) + acetic anhydride (E+ O E+) $\xrightarrow{H_2SO_4}$

FAQ: What if I can't determine which the nucleophile is or which is the electrophile?

Answer: The first backup plan is look at resonance structures of each. Resonance is a very powerful tool that can help us determine where partial negative and partial positive charges may reside. If that doesn't work, skip to step 2 (moving the acid) which might help you see it a little better.

FAQ: What about radicals? Are they nucleophiles or electrophiles?

Answer: Radicals are electron deficient species; however they are not true electrophiles in the classic sense. Please see the Overtime section ("What Not to do with EASE") for how to deal with these.

Exercise 1.1 Identify the Nu or E+ on the following molecules.

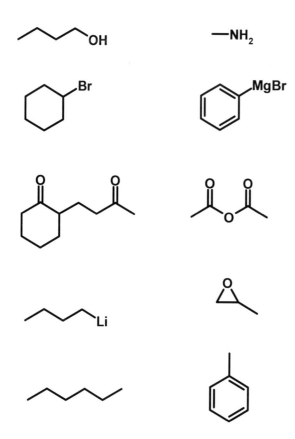

Exercise 1.2 Identify the Nu or E+ on the following molecules.

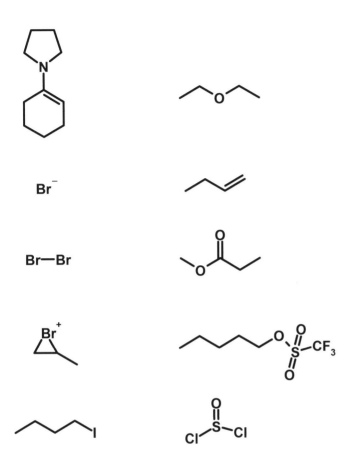

Step 2--"A" is for Acid: Identify any acids or bases and move the protons.

Acids

Traditional acids, also called "Bronsted acids" or "Bronsted-Lowry acids," will donate a proton (H⁺) to another functional group. Once a proton has been donated, the acid (without the proton) is referred to as the conjugate base.

[Reaction scheme: Acid (carboxylic acid, R-COOH) + Base (R-NH₂) ⇌ Conjugate Base (R-COO⁻) + Conjugate Acid (R-NH₃⁺)]

With respect to arrows, the curved, "electron flow" arrows will always point to a proton on the acid because the base possesses the electrons which are attacking the proton.

Further, you will notice that we used a two-way double-headed reaction arrow to symbolize the equilibrium which occurs between the compounds on either side of the reaction arrows. In other words, although most of the time the acid will have given its proton to the base, there is some percentage of the reaction mixture that has the conjugate acid giving up a proton to the conjugate base, thereby re-creating the starting materials and reversing the reaction.

FAQ: Will acid-base reactions always use a double-headed "equilibrium" arrow?

Answer: No. If your acid (or base) is very strong, for all practical purposes, the equilibrium will so greatly favor the products that you can use a one-sided "normal" reaction arrow, pointing to the right.

It now becomes a useful tool to understand acid strength. Think of this as how strong your football lineman is. While there are many ways to measure this strength, below is a chart that tracks acidity in DMSO, a common aprotic solvent. The key is to know that the smaller (or more negative) the acid's pKa, the stronger an acid it will be, and the easier it will give up its proton.

Compound	pKa	Compound	pKa
Hydroiodic acid	-10	Water	15.74
Hydrobromic acid	-9	Ethanol	16
Hydrochloric acid	-8	Acetone	19.3
Sulfuric acid	-3	Ethyl acetate	25
Hydronium ion	-1.74	Ammonia	36
Nitric acid	-1.5	Diisopropylamine	40
Trifluoroacetic acid	0.5	Toluene	41
Acetic acid	4.8	Benzene	43
Ammonium ion	9.3	Propene	43
Hydrogen cyanide	9.3	Ethene	44
Phenol	9.9	Methane	60

Another takeaway message from this chart would be to understand that certain classes of acids are stronger than others. For example, the hydrogen halides (HI, HBr, HCl, and HF) are much stronger than the carboxylic acids (i.e. acetic acid, trifluoroacetic acid). The absolute weakest acids are alkanes, which never really want to give up a proton so they very rarely act as an acid.

To elaborate on our first point, a common problem we see students constantly running into is that they do not readily recognize strong acids. This is a terrible mistake and should never happen; you will need to quickly recognize strong acids and understand which atom in the reaction will accept their proton. As far as strong acids go, you should immediately recognize the Front Seven:

STRONG ACIDS: HI, HCl, HBr, HNO_3, H_2SO_4, H_3PO_4, $HClO_3$

Just remember that it's the three hydrogen halides (HCl, HBr, HI) and the four acids that have N, S, P and Cl as central atoms. Many students remember the other four acids with the mnemonic, something you would never say to a girlfriend: "**N**ever **S**ay '**P**lease **Cl**ean'."

Also, as a rule, the more electron-withdrawing groups you place on an acid, the stronger that acid will become. This is because those electron withdrawing groups will help to stabilize the negative charge left behind once the proton is gone. This is easily depicted by comparing weaker acetic acid (pKa 4.8) to the stronger trifluoroacetic acid (pKa 0.5). Those three fluorine atoms on the trifluoroacetic acid greatly increase the overall acidity of the compound.

Weaker Acid (pKa 4.8)   **Stronger Acid** (pKa 0.5)

Acids from Unlikely Places:

You have now seen all of the All-Star acids. These are the strong acids which you will see again and again. They are the acids you know and the acids you love. However, if these acids are going to start the game for us, are there any acids that can come off the bench? The answer is yes. These are acids in disguise. Protons that you might not think are acidic, but can be abstracted easily in the presence of certain bases.

Protons which are alpha to a carbonyl can be acidic. As we show above, the proton is abstracted by base, leaving a carbanion behind. This carbanion is resonance stabilized, which helps make the proton more acidic. Once again, resonance has shown us what is likely to react.

So if one carbonyl makes a proton acidic, then two must make it really acidic. Welcome beta-diketones into the game.

Just as with regular carbonyls, beta-diketones will stabilize the carbanion formed, but this time it is stabilized by two carbonyls, which makes it even more acidic and even easier to remove.

Finally, one more unlikely place that you might find an acidic proton is on the end of an alkyne. The terminal proton on an alkyne has a pKa of approximately 25. This means it can be removed with a relatively strong base such as n-butyl lithium or sodium amide. The newly-formed carbanion is a great nucleophile, but beware, as with many other nucleophiles, it can also be a strong base once formed. This concern is magnified specifically with alkyne carbanions, as their linearity can make them a very non-hindered base able to pluck acidic protons relatively easily.

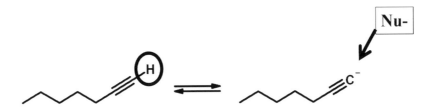

Take home points on acids:

1) Know that acids donate protons.
2) Know that the smaller (or more negative) a pKa is, the stronger the acid is.
3) Be able to recognize which classes of acids are stronger than others.
4) Know the "Front Seven" of acids.
5) Remember that electron withdrawing groups will make an acid stronger.

Lewis Acids:

A Lewis acid is an acid that accepts electron density. This is accomplished because there is a vacant, low energy orbital which can accept a lone pair of electrons. This definition of an acid is much broader than the Bronsted-Lowry definition and is very important in organic chemistry.

Some examples of Lewis Acids are shown below:

$AlCl_3$ BF_3 $ZnCl_2$ $TiCl_4$ $FeCl_4$ $BeCl_2$ Cu^{+2} Cr^{+3} Li^+ Mg^{+2}

It is important to note that like their Bronsted cousins, Lewis acids are never going to be negative, as you can't accept electrons if you are already negative to start with.

Lewis acids perform two major functions in undergraduate organic chemistry. The first of these is that they can combine with Lewis bases to form a Lewis acid-base complex. This is as simple as adding A + B, literally. The Lewis acid will accept electron density from the Lewis base, which will donate it. The two form a new bond and are now a complex, as shown below:

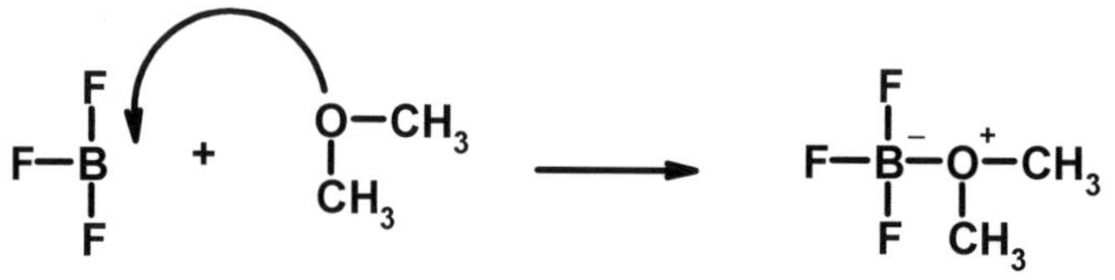

Boron Trifluoride
Lewis Acid

Diethyl Ether
Lewis Base

Borontrifluoride Etherate
aka BF3-Etherate
Lewis Acid-Base Complex

FAQ: Once you form a Lewis acid-base complex, can the Lewis acid still act like a Lewis acid?

Answer: Yes. In the example above, BF_3-etherate can still be used in organic chemistry reactions as a Lewis acid catalyst.

Second, and much more important, Lewis acids serve to "activate" certain electrophiles, making them more susceptible to attack by a nucleophile and thus more reactive. As stated above, Lewis acids accept electron density. But it is even more insidious than that as Lewis acids are like little burglars. They not only accept electron density from an electrophile, they are always trying to swipe someone else's because they are lacking it themselves.

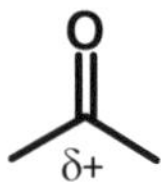

Somewhat electrophilic carbon

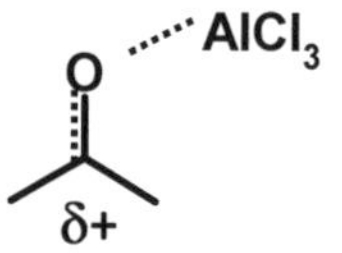

Much more electrophilic carbon

Organic chemists can use this to their advantage, in that Lewis acids can be used to make electrophiles even more electrophilic. Because Lewis acids are electron-deficient, they will seek out electron-rich atoms. The electron-rich atom (the oxygen atom in the previous example) will then steal density from the closest atom to it (the carbon atom), which results in the carbon becoming much more electrophilic. This oxygen atom is "robbing Peter to pay Paul" which makes the carbonyl carbon more prone to attack by a nucleophile.

You have seen numerous examples of this in your organic chemistry class, as shown below:

Reaction	Example	Lewis Acid Complex
Aromatic Halogenation	benzene + Cl_2 / $AlCl_3$ → chlorobenzene	$Cl-Cl \cdots AlCl_3$ with $\delta+$ and $\delta-$
Freidel-Craft Acylation	benzene + acyl chloride / $AlCl_3$ → acetophenone	acyl chloride with $O \cdots AlCl_3$ ($\delta-$) and $\delta+$ at C-Cl
Freidel-Craft Alkylation	benzene + CH_3Cl / $AlCl_3$ → toluene	$H_3C-Cl \cdots AlCl_3$ with $\delta+$
$LiAlH_4$ Reductions	aldehyde + $LiAlH_4$ → alcohol (OH)	R-C(=O)-H with $O \cdots AlH_3$, $\delta+$ at C, H^- attacking

Please take note: Lewis acids will manipulate your electrophile in more than one way. As shown, it can suck electron density away from your electrophile OR it can rip a halogen right off. In the latter mode, you will form a carbocation, as is the case for the Freidel-Craft alkylation and acylation. The difference between these two Freidel-Craft reactions is that the alkylation forms a standard carbocation which will rearrange if possible, whereas the acylation will form an acylium ion which will not rearrange. Both are very active electrophiles, but the rearrangement is a key difference between the two.

H₃C—Cl------AlCl₃ ⟶ H₃C+ + AlCl₄⁻   Carbocation, rearrangement can occur

[acetyl chloride---AlCl₃ complex] ⟶ H₃C—C≡O⁺ + AlCl₄⁻   Acylium ion, rearrangement can not occur

One way to think about this is in terms of bullies. Let's take the AlCl₃ example. The Lewis acid is the biggest bully and steals lunch money from the oxygen. The oxygen is a bit of a bully himself, though. The oxygen still wants to eat and steals some amount of lunch money from the carbon. Carbon is not a bully at all, but is actually a very sweet little kid. Carbon doesn't steal from anyone and is just left with no lunch money. Thus the carbon is now "poor," and more electrophilic than before it was subjected to the Lewis acid's bullying.

By the way, if you draw Lewis acids doing this on a sheet of paper it is traditional bullying, but if you draw if using ChemDraw on your computer it is considered cyber bullying. (Bad joke, sorry)

Take home points on Lewis acids:

1) They will complex with Lewis bases to form Lewis acid-base complexes.
2) They will activate electrophiles to make them more electrophilic.

Bases:

Bases, by the Bronsted-Lowry definition, will accept protons from acids. More accurately, bases will abstract a proton from an acid. Once the proton has been accepted, the base becomes its conjugate acid.

(CH₃)₃N + HCN ⇌ (CH₃)₃NH⁺ + CN⁻

Base   Acid   Conjugate Acid   Conjugate Base

As shown above, triethylamine will abstract a proton and act as a base. Once it is protonated, it is now its conjugate acid. As with any acid-base reaction, this is a reversible reaction, as the triethylamine can lose the proton and go back to being a simple base.

There are several different classes of strong/common bases that we will look at in this section.

Amines	Alkyl	Alkoxides	Hydroxides	Amides
$H_3C-N(CH_3)-CH_3$	$CH_3CH_2CH_2CH_2Li$	$CH_3CH_2CH_2ONa$	NaOH	$NaNH_2$

The first thing that should stand out to you when you look at the figure above is that all of the bases, with the exception of the amines, are complexed with group I metals (like sodium, lithium, or potassium) or group II metals (such as calcium or barium). The exception to this observation is an amine, which has a lone pair and is capable of accepting a proton to form a positive complex. The second thing that should stand out to you is that strong bases can have a negative charge on a central atom. Whether the central atom is carbon, oxygen, or nitrogen is immaterial, but what is important is that it will gobble up a proton from wherever it can to satisfy that negative charge.

A chart which we have found to be very helpful can be found on the next page. It contains some common bases, their pKa's, and which types of protons they are likely to be able to remove. The take home message from this chart is that the pKa of your base must be **_LARGER_** than the pKa of the acidic proton you are trying to remove for it to be successful. Base strength increases as you move down this chart.

Base	pKa	Will remove the following protons
NH3	9.2	carboxylic acid COOH (pKa 3-5)
CH3NH2	10.6	between dione (pKa 9), carboxylic acid COOH (pKa 3-5)
NaOH	16	phenol (pKa 10), between dione (pKa 9), carboxylic acid COOH (pKa 3-5)
NaOEt	16	phenol (pKa 10), between dione (pKa 9), carboxylic acid COOH (pKa 3-5)
NaO(CH3)3	18	phenol (pKa 10), between dione (pKa 9), carboxylic acid COOH (pKa 3-5)
HC≡C-	25	alpha to C=O (pKa 19-25), ROH (pKa 15-17), phenol (pKa 10), between dione (pKa 9), carboxylic acid COOH (pKa 3-5)
H-	35	alkyne (pKa 25), alpha to C=O (pKa 19-25), ROH (pKa 15-17), phenol (pKa 10), between dione (pKa 9), carboxylic acid COOH (pKa 3-5)
NaNH2	35	alkyne (pKa 25), alpha to C=O (pKa 19-25), ROH (pKa 15-17), phenol (pKa 10), between dione (pKa 9), carboxylic acid COOH (pKa 3-5)
nBuLi	50	NH3 (pKa 35), alkyne (pKa 25), alpha to C=O (pKa 19-25), ROH (pKa 15-17), phenol (pKa 10), between dione (pKa 9), carboxylic acid COOH (pKa 3-5)

In addition to the strong bases are numerous weaker bases you will encounter. Some of these are shown below.

$NaHCO_3$ $Zn(OH)_2$ NH_4OH

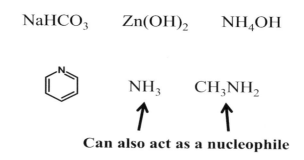

Can also act as a nucleophile

Like the strong bases, these weak bases will pick up a proton, however they can't just pick up any proton. Since these bases are weaker, the proton they abstract must be that much more acidic for the acid-base reaction to occur. This is where the pKa of your acid and base will be important. However, generally these are more likely to act as a "proton sponge" soaking up unwanted H^+ byproducts.

Further, there are a number of bases that, when speaking in a steric sense, are very, very large. These "Basic Big Boys" are the baddest bulkies on the field. Because of their size and bulk, they cannot react at most electrophilic sites on other molecules. This means that they can only act as a base, because they are too large to act as a nucleophile.

Below are the four most common bulky bases:

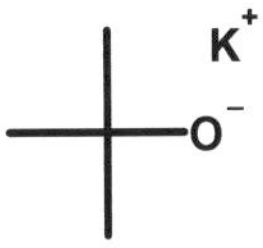

Potassium tert-butoxide

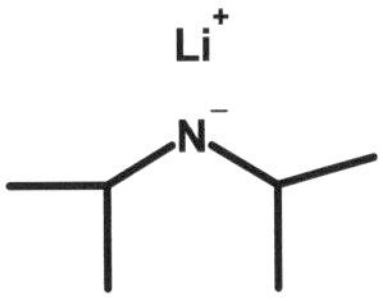

Lithium Di-isopropyl amide (LDA)

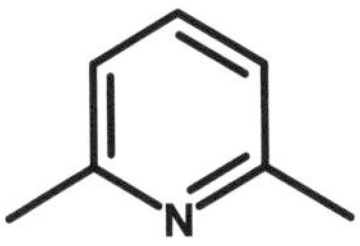

2,6 Dimethyl Pyridine

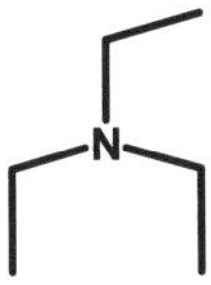

Triethylamine

In substitution/elimination problems, elimination will always be favored over substitution when using these bases, as it is impossible for them to act as a nucleophile.

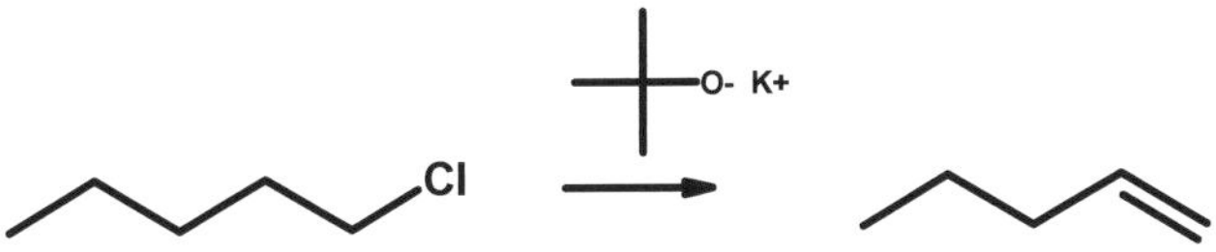

ONLY E2, NO SN2

Take home points on bases:

1) Bases suck up protons.
2) Many strong bases will have a negative charge.
3) Weak bases are not as good at abstracting protons, act as a proton sponge.
4) Bulky bases can only act as a base, not as a nucleophile.

Exercise 2.1 Identify the most acidic proton on the following molecules.

Exercise 2.2 You have four bases: t-butyllithium, sodium amide, potassium hydroxide, and ammonia. Which of these four bases can deprotonate the most acidic proton on the following molecules? Hint: there may be more than one answer.

The Bases:

BuLi $NaNH_2$ KOH NH_3

The molecules with acidic protons:

Exercise 2.3 Identify the following molecules as an acid or base. Then, identify whether it is a strong or weak acid/base.

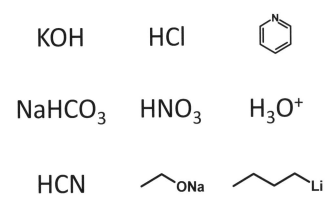

Step 3 -- "S" is for Sterics: Identify any steric impedance.

Steric hindrance is that fat bouncer at your favorite bar.

Steric impedance (aka steric hindrance) occurs anytime the size of your molecule gets in the way of your potential reaction. There are many analogies out there for steric impedance; the best one I've seen recently compares it to a big fat goalie. The goalie takes up the entire area you're trying to get to, and the bigger he is, the harder it is to get there.

Steric interactions can occur at one of two places: either on the nucleophile or on the electrophile. Many of you are used to looking for it on the electrophile. The best example of this is with SN2 substrates.

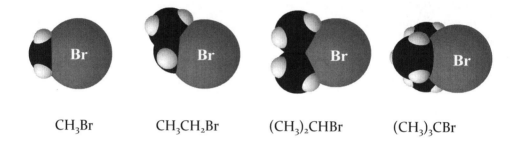

CH_3Br CH_3CH_2Br $(CH_3)_2CHBr$ $(CH_3)_3CBr$

In the above figure, a gray atom represents bromine. This representation makes it easier to see that the more methyl groups that are on the molecule, the more they hinder the backside attack to substitute for bromine. This is the easiest and one of the most clear cut examples of steric interaction, where the electrophilic carbon is blocked by a bulky group(s).

Less intuitive however, is when the bulky group blocks the nucleophile from reacting. The best example of this is with bulky bases. There are a number of bases that, when speaking in a steric sense, are very, very large. These "Basic Big Boys" are the baddest bulkies on the block. Because of their size and bulk, they cannot react at most electrophilic sites on another molecule. This means they can only act as a base, because they are too large to act as a nucleophile.

To repeat part of Step 2, the four most common bulky bases are:

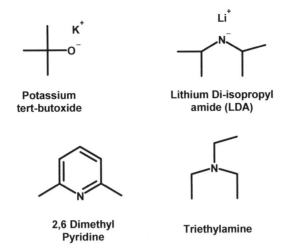

| Potassium tert-butoxide | Lithium Di-isopropyl amide (LDA) |

| 2,6 Dimethyl Pyridine | Triethylamine |

In substitution/elimination problems, elimination will always be favored over substitution when using these bases, as it is impossible for them to act as a nucleophile. They are just too big to be nucleophiles, yet they aren't too big to be bases, as a proton is much smaller and easier for them to get.

ONLY E2, NO SN2

<u>FAQ</u>: Can a bulky base *ever* be a nucleophile?

<u>Answer</u>: Only in very, very limited and specific cases, so if you see a bulky base is almost always going to be just a base.

Now that we've had a chance to see where steric interactions may occur, we can look at problems and decide what the best way to deal with any steric bulk is.

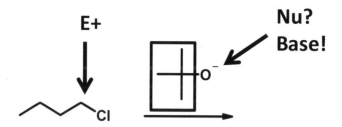

E- Step 1: Nucleophile might be a base, alkyl halide is E+
A- Step 2: There is a base, where is the acid?
S- Step 3: There IS a steric issue

The problem above is a typical E2-type reaction. In step 1, we identify the electrophile easily but are unsure if the oxygen anion is our nucleophile. So we can tentatively label it as "Nu?", and put a question mark there just in case we change our mind later. In step 2, we identify the potential base very quickly. Hence, after the first two steps we have an idea but we are not totally sure if the alkoxide is going to act as a base or nucleophile.

In step 3, we look for steric interactions, where we are able to see that the alkoxide will act as a base and not as a nucleophile. The large t-butyl group attached to the alkoxide is going to preclude this molecule from acting as a nucleophile. We label steric bulk by placing a box around the bulky group.

<u>FAQ</u>: Are t-butyl groups a good hint that there will be steric interactions?

<u>Answer</u>: Why yes, that's an excellent observation. Either that, or it is an SN2/E2 reaction. Helmet sticker for you.

We can now identify several groups which can hinder organic reactions. Two of them are relatively obvious; one of them is not.

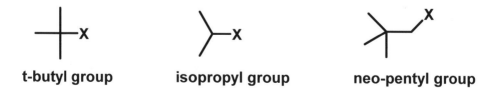

The obvious two groups are the t-butyl group and the isopropyl group. They are so large, especially the t-butyl, that they can cause major havoc in your organic chemistry reaction. The less obvious one is the neo-pentyl group. The carbon that our electrophile is attached to is a primary carbon, but the large adjacent quaternary carbon is enough to disrupt a reaction.

Even less obvious are ring and bridging structures where certain electrophiles are blocked.

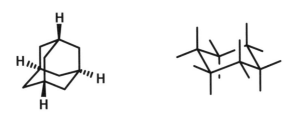

So congratulations, you have found a sterically hindering group in your reaction, now what? One of three outcomes will occur: 1) the bulky group will keep ANY reaction from occurring, 2) the bulky group will stabilize an intermediate (ala SN_1/E_1 carbocations), and allow you designate a new, more powerful electrophile, or 3) you will ignore the bulky group and proceed with the mechanism because the reactants can overcome the steric bulk.

FAQ: Ok, I know I have a bulky group in the reaction. How do I know which of these outcomes will occur?

Answer: This is a very difficult question and will partially depend on the other components of your reaction. If you have a great leaving group and a polar protic solvent, you are likely to form a carbocation. If you have a small nucleophile and/or a strong base, you are likely to proceed with the reaction, possibly after the carbocation has formed. If you have a sterically hindered nucleophile, generally the reaction will have difficulty proceeding. If you are doing a cyclization reaction, the reaction may have great difficulty with steric bulk. Overall, this can become a judgment call and might require some notes next to your mechanism to convey to your professor what you were thinking. One good thing is that professors *usually* won't have a reaction that gives no product due to steric bulk on an exam unless it is an example they have presented before.

FAQ: Is there ever a point where steric hindrance can stop a reaction from occurring altogether?

Answer: Yes, there is. Here are three instances where sterics will completely stop a reaction from happening. The first is a Diels-Alder reaction where there are t-butyl groups on the 1,4 positions. The second is a hindered and tertiary substrate trying to react under SN1 optimized conditions. The third is any reaction that your professor tells you won't work because of steric interactions. The first two of these examples are shown below.

Poor Yield

NaCN
Acetone
NO REACTION

This ball-and-stick model of the second reaction helps show why it doesn't work. Notice how many carbon atoms are between the nucleophile and the electrophilic carbon.

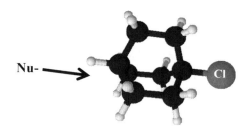

Exercise 3.1: Rank in order of increasing nucleophilicity.

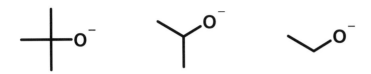

Exercise 3.2: Which is the following has the least amount of steric hindrance at the electrophilic carbon?

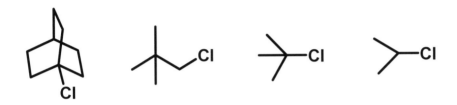

Step 4 --"E" is for Electron Flow: Move your electrons from nucleophile to electrophile.

We've waited long enough. We've trained, practiced, and learned everything we need to know about our players. We've put ourselves in a good position to know exactly what is going to happen in the reaction. Now we're ready for the magic: making the electrons flow.

<u>Rule of the game</u>: Electrons always always always always flow from nucleophile to electrophile. [I said it four times therefore it must be important].

<u>Rule of the game</u>: Arrows show electron flow, and the arrowhead points to where the electrons are going.

Let's start with a relatively simple example:

      E- Step 1: Both the E+ and the Nu- are relatively obvious
      A - Step 2: Grignard can be bases, but there is not an acid
      S - Step 3: There are no steric issues
      E - Step 4: Move electrons from the Nu to the E+

In step 1, we can identify the nucleophile and the electrophile quickly. Grignards will always be nucleophiles and carbon dioxide is an excellent electrophile. In step 2, we recognize that the Grignard could also be a base, but there is really no corresponding acid, so we'll ignore this step for now. In step 3 we see there are no steric issues, which brings us to step 4 where the electrons from our nucleophile will attack the electrophile. In this example, because our electrophilic carbon already has four bonds, we must break a bond to accommodate the new one that we are forming. This is easy because we have a double bond that we can change to a single bond. Because our electron flow ends at the oxygen we end up with a negative charge on that atom, which forms a salt with $MgBr^+$. Our reaction is now complete, pending an acid work up to give the carboxylic acid as a final product.

If you were to talk your way through organic chemistry mechanisms like we did above, there will be very few mechanisms that you won't be able to reason your way through.

Now let's try an example that is a little more complex, but not too difficult:

On its face, this appears to be a Diels-Alder reaction, but there are still a couple of questions which pop out to us. Which end of the diene is more strongly nucleophilic? What part does the Lewis acid play? Could there be any side reactions?

As we start to look at this problem with the EASE method, we see there really isn't any part of either molecule that stands out as a nucleophile. Thus, our next attempt should be to try creating different resonance structures to find a nucleophile.

Possible resonance structures:

After creating several resonance structures, the problem becomes much clearer. We can more clearly see that there is one position which stands out as a nucleophile and three which are candidates for the electrophile. Let's start our EASE process and see where it goes:

E- Step 1: One good Nu candidate, three decent E+ candidates
A- Step 2: LA present, will coordinate with O of C=O
S- Step 3: No steric issues
E- Step 4: Attack the 1,4 position with Nu

In step 1, we have identified the only truly nucleophilic site, which is the terminal end of the diene, closer to the ether. This determination is supported by the resonance structures we previously drew. With respect to the electrophile, we have three possible sites. For now, let's label them all and come back to determine the most electrophilic later. In step 2, we have a Lewis acid, which will coordinate to the most Lewis basic site, which in this case is the carbonyl oxygen, as it is more basic than the ether or ester oxygen. Because we have a Lewis acid coordinated to the carbonyl, it is going to draw electron density away from the carbonyl carbon, which in turn will take it from the double bond which is alpha,beta to the carbonyl. This is the "cascading bully" theory, where each group robs electron density from another group close to it, until you reach the end of the line and there is nobody left to steal from. The net result is that 1,4 position becomes much more electrophilic than everyone else, and now can be labeled as our most electrophilic site. Step 3 is quick as there are no steric issues, so we can move right on to step 4. In step 4, the most nucleophilic site (terminal double bond closer to ether) attacks the most electrophilic site (1,4 to the carbonyl) and the electrons flow from there to give the final product which is the cyclohexene shown.

FAQ: Could the diene have attacked to 1,2 position on the carbonyl?

Answer: This is not a favorable reaction, as soft nucleophiles and many carbanions are more likely to attack the soft 1,4 position. Steric issues at the carbonyl can also make this less favorable.

Now that you have a feel for what one game is like, let's try a doubleheader (multistep problem):

The problem above is the most difficult that we have seen yet. There are several carbonyls, no clear acid or nucleophiles and possible steric considerations. This may also end up being a multistep problem, adding to the complexity. However, if we continue to play the game the same way we have been doing it all along, we can still win.

E - Step 1: Three C=O and one C=C could all be E+, no real Nu-
A - Step 2: Base is obvious, 2 acidic protons, deprotonate now
S - Step 3: No real steric concerns
E - Step 4: Nu attacks 1,4 to carbonyl

Step 1 has three carbonyls and an alkene which are all potential electrophiles. Our nucleophile is not obvious at this point, but a carbanion between those two carbonyls looks like a pretty good candidate. Step 2 has no true acid, but shows sodium ethoxide as a base. This base is definitely strong enough to deprotonate the carbon between the carbonyls, but not strong enough to deprotonate anything else. That means we should deprotonate the carbon now and call it our nucleophile. Step 3 shows no real steric concerns, so step 4 become much clearer. Because the 1,4 position is "softer" than the 1,2 position, a "soft" nucleophile (like a carbanion) is more likely to attack through a conjugate addition to the unsaturated carbonyl, as shown. [For more information, on hard and soft nucleophiles and electrophiles, please consult your textbook]. Our two steps ended up being deprotonation followed by nucleophilic attack.

The final way we would like to highlight how these reactions can occur is through the intramolecular pathway. This means the nucleophile and the electrophile are on the same molecule and form a ring as the new product. Two simple examples of this are shown below. In the first reaction, an alcohol attacks a carboxylic in the presence of an acid catalyst present to form a new ring, which in this case is a lactone (aka cyclic ester). This reaction is favored because water is driven off as a byproduct and because an unstrained six member ring is formed.

The second example is a bit more complicated. Here, we show strong base KOH, promoting the attack of an alcohol on an alkyl halide. This reaction occurs because the nucleophile is forced to be close to the electrophile and the reaction conditions are dilute so that intermolecular reactions are limited. Hence, even though the three-membered ring (epoxide) formed is very strained, the reaction proceeds because it has been forced to through manipulating the conditions.

Exercise 4.1 What is the major product(s) of these reactions?

H—≡—/ $\xrightarrow{\text{HBr}}$ [ ]

~Br $\xrightarrow{\text{NaCN}}$ [ ]

cyclopentyl-OH $\xrightarrow{\text{cat. HCl}}$ [ ]

~~≡—H $\xrightarrow[\text{2) CH}_3\text{CH}_2\text{Br}]{\text{1) NaNH}_2}$ [ ]

$(CH_3)_2CuLi$ + ~~I $\longrightarrow$ [ ]

~⩘ $\xrightarrow{\text{Br}_2}$ [ ]

cyclohexadiene $\xrightarrow{\text{Excess HCl}}$ [ ] [ ]

H_2N~~~C(=O)CH$_3$ $\xrightarrow{\text{cat HCl}}$ [ ]

2nd Quarter:

7 Easy mechanisms you MUST know.
3 Intermediate mechanisms you should know.
7 More mechanisms all-stars will know.

Congratulations. You've made it through the warm-ups (introduction), you've learned all of your teammates (reagents), and you know the rules of the game (the EASE method and rules of chemical reactions). Now it's time to go to the second quarter of our game.

In this quarter, we'll examine some reactions that you must know like the back of your hand. Most people would define these as easy reactions, which means we need to know this part of the playbook inside and out.

WAIT! *Even though the first part of this quarter will focus on some pretty easy reactions, we would encourage you not to skip it. The whole idea of the EASE method is that you should be able to use this step-wise approach on almost any organic chemistry problem. Therefore, you should practice it on the easiest problems you can first, before moving on to more difficult circumstances. We will also point out some nuances of organic chemistry reactions in this quarter that you will need to know for tougher problems.*

Need to Know Mechanism #1: SN1

Generic:

Nu- + R—LG ⟶ R—Nu + LG-

Example:

><Br + HOH ⟶ ><OH + HBr

This reaction, in its purest form, is substituting a nucleophile for a leaving group. It is also one of the most basic reactions of the first semester course. The reaction is very versatile and can be accomplished on any number of substrates which are capable of stabilizing the carbocation intermediate.

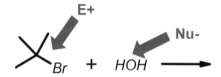

E- Step 1: E⁺ is attached to the LG (Br), water is Nu⁻
A- Step 2: No acid or bases present
S- Step 3: Yes there is a steric issue

Since this is our first mechanism demonstration, we will go a little slower here. Let's examine the top portion of the figure above. In step 1, we must identify the electrophile and nucleophile. As we remember from the 1st Quarter, the carbon attached to the halogen (an alkyl halide) is almost always electrophilic, and water is a good nucleophile. On your exam, clearly mark these with an "E+" and a "Nu-," with arrows pointing to the exact spot. In step 2 we look for any acids and/or bases, and move the proton if they are strong. In this problem, there are no acids or bases, so we can continue on to the next step. In step 3, we look for potential steric issues and pause if they are present. In this example, we have one of the most obvious steric hindrances, a tert-butyl group. Because we have this t-butyl group, we necessarily have a tertiary carbon. We must address the t-butyl group, which we can do in one of three ways: stopping the reaction, creating a carbocation or ignoring it. In this case, we can detach the leaving group to make a tertiary carbocation. As soon as we create a carbocation, we must rearrange it, if possible, which in this case it is not. Once we have done this, we have a new electrophile in the reaction, so it makes sense to start the process over.

Examining the figure below now, we see that for step 1 ("E") we have a carbocation electrophile and still have the water nucleophile. There are no acids or bases ("A"), and the steric issues ("S") are lessened because now we have a carbocation. This means we can move to step 4 ("E") and have the nucleophile attack the electrophile, giving us the final product.

$$\overset{\text{E+}}{\downarrow}\quad \overset{\text{Nu-}}{\searrow}$$

$$\underset{\text{Br}}{\diagup\!\!\!\diagdown} + HOH \longrightarrow \overset{+}{C} + HOH \longrightarrow \boxed{\underset{\text{OH}}{\diagup\!\!\!\diagdown}} + HBr$$

E- Step 1: E⁺ is carbocation, water is still Nu⁻
A- Step 2: No acid or bases present
S- Step 3: Lesser steric issues
E- Step 4: Nu attacks E⁺

While this may seem like a lot of work to go through for a simple SN1 reaction, we must remember that our goal for the EASE method is not only to use it on simple mechanisms like this, but to be able to apply it to more difficult mechanisms that we might never have seen before. So this is great practice for us.

FAQ: Where in the EASE method is the best place to think about carbocation rearrangement?

Answer: Carbocation rearrangements are one of those places that professors love to try to throw a curve ball at students. In the EASE method, you need to think about carbocation rearrangements as soon as the cation is formed. The important thing to remember about carbocations is that if it can rearrange, it WILL rearrange.

Need to Know Mechanism #2: SN2

Generic:

Nu- + R—LG ⟶ R—Nu + LG-

Example:

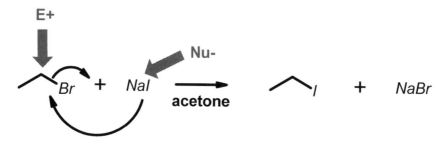

As with the previous mechanism, this mechanism substitutes a nucleophile for a leaving group. However unlike SN1, this reaction proceeds in one step, concerted fashion with the nucleophile "pushing off" the leaving group.

E- Step 1: E+ is attached to LG (Br), I is the Nu-
A- Step 2: No acids or bases present
S- Step 3: No Steric Issues
E- Step 4: Nu- attacks E+

This should be one of the more simple mechanisms to sort through. As with many of these types of reactions, the alkyl halide is the electrophile; iodine is the nucleophile. In step 2, we see that there are no acids or bases present, and there are no steric issues. Therefore, in step 4, the nucleophile attacks the electrophile at the electrophilic carbon and pushes off the leaving group, giving the product shown. [Note: This specific reaction works because NaI is soluble in acetone, while NaBr is not. Think LeChatlier's principle.]

FAQ: What if the electrophile had some steric bulk to it?

Answer: If the electrophile or the nucleophile in SN2-type reaction is sterically hindered, then the reaction will favor elimination instead of substitution.

FAQ: What types of solvents promote SN2 reactions?

Answer: Polar aprotic solvents, such as acetone, DMF, and acetonitrile. Look for these as clues that you have an SN2 reaction.

FAQ: Why is the degree of substitution of electrophile (starting material) important?

Answer: SN2 reactions only proceed when the nucleophile has room to attack the electrophile. Therefore, if the nucleophile can't get at the electrophile there is no reaction. Below is a "to scale" representation of the steric hindrance caused by different alkyl halides. This clearly shows how difficult it can be to attack any hindered alkyl halide.

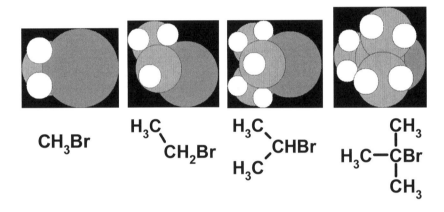

FAQ: What does the SN2 transition state look like?

Answer: The alkyl halide becomes planar, and as the nucleophile gets closer, the leaving group gets further way from the electrophilic carbon. One way to think about it is that the electrophilic carbon has three full bonds that don't change, and two partial bonds that weaken and strengthen as the nucleophile moves closer.

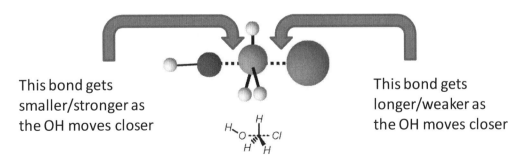

This bond gets smaller/stronger as the OH moves closer

This bond gets longer/weaker as the OH moves closer

Need to Know Mechanism #3: E1

Generic:

Base + R—LG ⟶ R=R

Example:

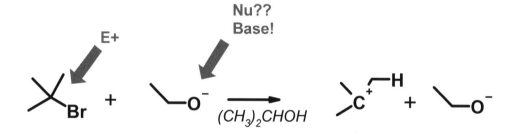

For every yin there has to be a yang. This is the case between SN1 and E1 reactions. Both proceed through a carbocation mechanism, but an SN1 reaction gives the substitution product, while an E1 reaction gives the elimination product. Elimination is favored at higher temperatures and with non-nucleophilic (aka bulky) bases. The net result of the reaction is that the alkyl halide is converted into an alkene.

E- Step 1: E⁺ is attached to the LG (Br), O⁻ is possible Nu⁻
A- Step 2: O⁻ is definitely a base
S- Step 3: Yes there is a steric issue

Please refer to the above reaction. For step 1, it's easy to determine that the electrophile is our alkyl halide. The oxygen anion, also called an alkoxide, could be a nucleophile in very limited circumstances but is always a strong hindered base. For step 2, we have already determined that the alkoxide is a strong base. For step 3, we see that there is a steric issue: the t-butyl group on our alkyl halide. At this point we stop the EASE method because these steric considerations must be addressed. As with the SN1 reaction, the first thing that takes place is formation of the carbocation. Once this is formed we go through steps one through four again.

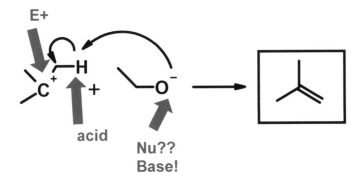

E- Step 1: E⁺ is carbocation, unsure about Nu⁻
A- Step 2: O⁻ is definitely a base, H could be an acid
S- Step 3: Yes there is a steric issue
E- Step 4: Have base abstract proton, and move electrons

For step 1, we see that the carbocation will definitely be our electrophile, but we're still unsure about the nucleophile. For step 2, the alkoxide is still a strong base which means there must be an acidic proton for it to react with. That acidic proton is one carbon away from our carbocation. While there are still steric issues associated with the reaction, it's clear what happens from here. The base abstracts the proton and we form our new alkene, thus completing the elimination reaction.

It is worth noting that this is one of the very few reactions where step 4 does not involve a classic nucleophile attacking a classic electrophile. In this reaction, as with the E2 reaction, the mechanism ends with an acid/base reaction. While in the strictest sense of the words, acids and bases are actually electrophiles and nucleophiles, respectively, it is not what we are used to seeing in organic chemistry reactions. This is a rarity. A clue to this reaction pathway is that you could not identify any good nucleophile in the reaction mixture.

FAQ: What clues exist which would help me to know that this is an E1 and not a substitution?

Answer: There are two clues here which would tell us this is an E1 reaction. The first is that we have a strong and hindered base, which cannot be used as a nucleophile. The second is that we have a polar protic solvent that will help stabilize a carbocation. If it were shown, another clue might be if the temperature of the reaction were high.

FAQ: My professor has discussed the Hoffman product of elimination. What is that?

Answer: The Hoffman product is often referred to as the anti-Zeitsev product. While the Zeitsev product will give you the more substituted alkene, the Hoffman product will give you the less substituted alkene. There are two factors which will maximize the Hoffman product. The first is the bulkiness of the base, as hindered bases are more likely to give you a Hoffman product. The second is the leaving group, as fluorine and sulfonate leaving groups are more likely to give you Hoffman products. As a general rule, if you have a tertiary alkyl halide and t-butoxide as your base, you can get the Hoffman product as the major product.

	Zaitsev Product	Hoffmann Product
Reaction 1	71%	29%
Reaction 2	28%	72%

	Hoffman	Zaitzev
X = I	19%	81%
X = Cl	33%	67%
X = F	69%	21%
X = OTs	83%	17%

Need to Know Mechanism #4: E2

Generic:

Base + R—LG ⟶ R=R

Example:

CH₃CH₂CH₂Br + (CH₃)₂CHO⁻ → (in acetone)

The outcome of an E2 reaction is the same as that of an E1 reaction, but the pathway to get there is different. Whereas the E1 reaction was stepwise through a carbocation intermediate, the E2 reaction is concerted, meaning there is no intermediate.

[Mechanism diagram: E+ attached to Br on alkyl halide, acidic H on β-carbon, base (Nu??/Base!) is (CH₃)₂CHO⁻ in acetone, producing propene + (CH₃)₂CHOH]

- E- Step 1: E+ is attached to LG (Br), O- possible Nu-
- A- Step 2: O- is definitely a base, H on carbon #2 is acidic
- S- Step 3: Yes steric issues
- E- Step 4: Have base abtract proton and move electrons

In the above reaction, we can find the electrophile pretty quickly to satisfy step 1. The nucleophile is not so apparent, so let's skip it for now and move on. Step 2 becomes a bit clearer as we have a strong base present. This is a clue to us that we should now identify the most acidic proton in the reaction. That proton is the one beta to our alkyl halide. Step 3 might be the most important step in our process for this reaction because it's where we identify that we have a sterically hindered base. This is the key because it means that it will act as a base and not as a nucleophile. We can now proceed to step 4, where the base abstracts the most acidic proton and the elimination occurs giving the alkene product shown.

Again, like with the E1 reaction, this is one of the few instances where we won't have a true nucleophile attack, as the reaction is completed by proton abstraction and elimination.

FAQ: How do I know whether a reaction is going to be an elimination or substitution? And which mechanism will it go through?

Answer: Please see the chart below. We found this to be a great help for all organic chemistry students. It is a great guide for knowing which reaction will happen under what conditions.

	S_N1	$E1$	S_N2	$E2$
Reaction Mechanism	2-step with carbocation	2-step with carbocation	Concerted	Concerted
Strength of Nucleophile	Can be mediocre, must be non-basic	Can be mediocre, must be basic	Strong, non-basic, non-bulky	Strong and basic
Leaving Group Ability	Must be great	Must be great	Can be mediocre	Can be mediocre
Solvent	Polar protic	Polar protic	Polar aprotic	Polar aprotic
Product Stereochemistry	Racemic*	Alkene, bulky groups are trans to each other	Inversion of any chiral center	Alkene
Primary Substrate	No reaction	No reaction	Highly favored	Favored only with strong base
Secondary Substrate	Only with non-basic nucleophile or if carbocation is resonance stabilized	Only if carbocation is resonance stabilized (benzylic or allylic)	Need strong non-basic nucleophile, otherwise competes with $E2$	Favored with strong base
Tertiary Substrate	Favored with non-basic nucleophile	Competes with S_N1	No reaction	Can occur in strong base

Need to Know Mechanism #5: Electrophilic Aromatic Substitution

Generic:

[Benzene + E-Y, Lewis Acid → substituted benzene with E + H-Y]

Example:

[Toluene + propyl chloride, AlCl₃ → para-isopropyltoluene + HCl]

This mechanism is meant to encompass all of the EAS reactions, but we will illustrate it with a Friedel-Craft alkylation because it is good for demonstration purposes.

[Mechanism diagram showing Nu sites on toluene, E+ on alkyl chloride with AlCl₃ Lewis acid, formation of carbocation with rearrangement, plus AlCl₄⁻]

- E- Step 1: 3 potential Nu sites
- A- Step 2: AlCl3 is the Lewis Acid, takes Cl, rearrange
- S- Step 3: Some steric hinderance with methyl group, will go para
- E- Step 4: Electrons flow from Nu to E+

[Final step: Nu attacks E+ carbocation to form para-isopropyltoluene product]

In this reaction, step 1 labeling of nucleophilic and electrophilic is fairly obvious. In step 2, we determine there is a Lewis acid present. As with a Bronsted acid, we need to complex the acid to a base in this step. There are two potential sites for the Lewis acid to coordinate to, but the most logical site is the chlorine atom. Once this coordination occurs, it will cause the halogen to pop off, forming a carbocation. As with all carbocations, if they can rearrange, they will rearrange. Hence, we can now rearrange our carbocation to a more stable, final form of our electrophile. In step 3, we see that there are some minor steric issues which will cause the para site to be the preferred nucleophile. In the final step, the nucleophile attacks the electrophile, briefly causing the aryl ring to lose aromaticity. This is a highly unstable state which is made more stable once aromaticity is regenerated. This is accomplished by loss of a proton, clearing the way for the return of the ring's aromatic character.

FAQ: Can an EAS reaction proceed if I have strong electron withdrawing groups on the aryl ring?

Answer: Generally not. The nucleophile in this reaction is the aryl ring. Therefore, we want it to have all of the electron density it can, which electron withdrawing groups will only impede.

FAQ: Do all EAS reactions proceed through a cation?

Answer: No. However, the Friedel-Crafts alkylation and acylation do. Plus, because the Friedel-Craft alkylation proceeds through a carbocation, you can't forget about carbocation rearrangements.

Intermediate Mechanism #1: Nucleophilic Addition to a Carbonyl.

We will demonstrate this mechanism using an example where an amine attacks an activated carbonyl to form an amide. However, you should remember that this reaction can occur with a lot of nucleophiles, using a very similar mechanism. Additionally, we will go a little slower on this problem, to make sure you understand and are starting to master the EASE process.

Evaluation of this problem starts with step 1, where we quickly see an acid chloride as an electrophile and ammonia as a nucleophile. Both are pretty active, so we would expect this reaction to happen quickly. In step 2, we see that ammonia is also a weak base, but there is no acid really present. Hence, we will ignore step 2. In step 3, we notice a large aryl ring off of our electrophile. While we should take note of this, we can ultimately ignore it because ammonia is such a small nucleophile. Therefore, we can go right to step 4, where the nucleophile attacks the electrophile and forms a tetrahedral intermediate.

E-Step 1: E+ is acid chloride, Nu- is ammonia
A-Step 2: NH3 is a weak base, no acid present
S-Step 3: Phenyl ring should be noticed, put box around it
E-Step 4: Nu- attacks E+

Now that we have performed our first step in the process, let's reassess. The first thing that stands out here is that we have a negatively-charged oxygen and a positively-charged nitrogen. It makes sense to perform a proton transfer here in order to satisfy those charges. Now, our reaction mechanism becomes a little clearer. At this point, the remaining ammonia can act as a base and remove the proton from the OH, causing the carbonyl to regenerate itself and kick off chlorine, to produce the final product. This is referred to as dehydrohalogenation and occurs because chlorine is a better leaving group than the hydroxyl group or the amine.

FAQ: In this specific problem, does conjugation play any role in the regeneration of the carbonyl?

Answer: Yes it does. Because we have a phenyl ring there, the carbonyl which is regenerated is now in conjugation with the ring. This is a bit of an extra driving force towards regeneration. But while this is helpful and speeds up the reaction, it is not essential, as the regeneration would still occur without the conjugation.

Intermediate Mechanism #2: Alkene Addition

The second intermediate mechanism you should know is alkene addition, demonstrated here by an alkene reacting with bromine in water. When assessing step 1, we know an alkene can be a nucleophile and we know bromine can be an electrophile. Water is also a nucleophile, so let's go ahead and label those. There really isn't an acid or a base present here so we're not going to worry about step 2 and there are no steric issues. Hence, we have two nucleophiles and one electrophile. In this problem, the bromine will be attacked by the alkene, because there is no reaction between bromine and water without some sort of catalyst or extreme conditions.

The alkene attacks one bromine atom and ejects the other bromine atom, so we end up with the bromonium intermediate. At this point, we will reassess, as we know the bromonium intermediate is not stable and not our final product. For step 1, we know the two carbons attached to the positively-charged bromine are electrophilic. The water is still a nucleophile, as is Br⁻, which are both floating around in solution. Let's label them both as nucleophiles. After we go through step 1, it appears we have two electrophiles, two nucleophiles, no acid/bases, and no steric issues. In step 4, our nucleophile will attack the electrophile, which in this case is the water attacking the more substituted position on the bromonium intermediate. The reason for this attack is this electrophilic site on the bromonium intermediate is more capable of handling a cation, as it is more substituted. Greater substitution means greater stabilization, so water is going to attack there. After we move these electrons, we finish up with the product shown.

FAQ: Br⁻ is a better nucleophile than water, why doesn't Br⁻ attack to give the dibromide product?

Answer: The bromine could have attacked but it didn't because it didn't get the chance. Bromine is a much better nucleophile than water but there is only one mole of Br⁻ in the reaction mixture while there were many, many more of moles of water floating around it. The electrophilic site had a much slimmer chance of finding a bromine and it had an exponentially greater chance of finding water. So simple statistical probability says water is going to attack that bromonium intermediate, not Br⁻.

Intermediate #3: Dieckmann Reaction

The final intermediate mechanism we're going to talk about is the Dieckmann reaction, which is a good example of an intra-molecular cyclization. We're going to spend a little less time on this reaction because hopefully you are more comfortable with the EASE method at this point.

For step 1, we know esters are good electrophiles, but we don't know what the nucleophile is in this reaction. So label the carbonyls as electrophiles and keep your eyes open for a nucleophile. In step 2, the alpha protons are acidic and we know that sodium methoxide is a base. So we are going to deprotonate at the most acidic site, which is alpha to carbonyl, and reassess. Now, it's pretty obvious what happens here: we have one nucleophile and one electrophile, with the carbonyl as the electrophile and the deprotonated alpha carbon as a nucleophile. There are no steric issues so we can move directly to step 4. Here, the nucleophile attacks the electrophile, kicks off the ester portion of the carbonyl and cyclizes our ring, forming a cyclic ketone. Of note, is that we have also formed a beta-di-carbonyl, where the alpha carbon is now particularly acidic.

FAQ: Why do we have to use sodium ethoxide, could you use sodium methoxide?

Answer: No, you have to use sodium ethoxide. Remember that methoxide and ethoxide can act as nucleophiles. If we use sodium ethoxide, we are going to add ethoxide to the carbonyl and we're going to kick off ethoxide. So there is a reaction happening in solution but it doesn't matter because we are substituting for the same ester component. If we use methoxide then we would be mixing up our esters and created different products after the cyclization.

Using Ethoxide: ester + $^-OCH_2CH_3$ ⇌ ester + $^-OCH_2CH_3$ — Same product on both sides

Using methoxide: ester + $^-OCH_3$ ⇌ ester + ester + $^-OCH_2CH_3$ — A mixture of products

7 More Mechanisms All-Stars Will Know:

Because of constraints on space, we won't go into these mechanisms any more than we already have. But the point to be made here is an important one: The EASE method will work on all of these reactions. By applying the principles you have learned to this point, you should be able to determine the mechanism for any of the reactions below. Most of these reactions are ones you will see at some point during the semester, so you might as well get acquainted with them now.

Wittig

Grignard

Diels-Alder

Robinson Annelation

Aldol Reaction

Claissen

Fischer Esterification

3rd Quarter:

Putting it all Together

Easy Problem: What is the Mechanism?

$$2 \text{ CH}_3\text{CH}_2\text{OH} \xrightarrow[\text{Heat}]{H_2SO_4} \text{CH}_3\text{CH}_2\text{-O-CH}_2\text{CH}_3$$

Here is how this breaks down with the EASE method:

We have two moles of ethanol in the presence of sulfuric acid and heat. We're going to go a little faster with our explanation because you are more familiar with the EASE method now. Further, we don't need to write out each and every step. So let's get started with this mechanism. We have two moles of alcohol and catalytic sulfuric acid. The alcohol is a weak base and a nucleophile. H_2SO_4 is a strong acid, labeled with an "A", which is present, albeit in catalytic quantities. While we don't have a clear electrophile, we do have a strong acid. This means the oxygen atom can be the base, so we can transfer that proton and create a really good leaving group at the end of the ethanol. Remember, we still have another mole of ethanol floating around. After proton transfer, our nucleophile is the other mole of ethanol and our electrophile is the carbon attached to the leaving group (a protonated OH). The ethanol attacks that carbon and ejects water. Please note, this is **_not_** a reversible step. This is our permanent step where we've created a protonated ether and water. Finishing out the reaction is very simple now. All we do is lose that proton, which is a reversible step, and form diethyl ether as a product. In total, the reaction was three steps: a proton transfer from an acid, nucleophilic attack, and proton transfer/catalyst regeneration. Pretty easy.

Intermediate Problem: What is the Mechanism?

The first thing we want to notice in this problem is the word "then." This is a very clear instruction from your professor that you have two separate steps in the reaction. We have a strong acid followed by addition of methylamine. Because this is performed in two steps, we are going to ignore methylamine until the acid has reacted first. We can keep what the electrophile or nucleophile might be in the future in the back of our mind, but the oxygen is going to be a base and the HCl is an acid right now. This means we will protonate the oxygen first.

Remember, the first step is a proton transfer ("PT") and therefore is a reversible step. Thus, we get a positively charged oxygen on top of that carbonyl. Now we can do the full EASE method, which starts with identification of methylamine as a nucleophile and the protonated carbonyl as an electrophile (remember resonance here, which shows it is super-electrophilic). With no steric hindrance in sight, we can go ahead and have our nucleophile attack the electrophile. This is a reversible step because we can always go backwards by kicking off methylamine. After this step, we have methylamine attached to the central carbon, along with a protonated nitrogen. We can do another proton transfer because we know we need to lose the hydroxyl group somewhere. Protonation of the hydroxyl group makes water, which is an amazing leaving group. This is a permanent step, as we eject water and form the imine.

One last note on this reaction: one of the nice things about problems where you are asked to figure out the mechanism is that you already know what the product is. Since you know what product you need to get to, it is much easier to figure out a route to that product. In this problem, we had to kick off oxygen somewhere. The easiest way to do that is to kick off water as the leaving group.

Hard Problem: What is the Mechanism?

Our final mechanism example is a harder problem which will require several steps to show the final mechanism. As stated above, the nice thing about mechanism problems is that you already know the product of the reaction, therefore you just need figure out how to get there. In this case, we have a seven-carbon ring being combined with a four-carbon alpha-beta unsaturated ketone. The product is an 11-carbon fused ring, so it stands to reason that we must show the mechanism of how these two were combined.

This also provides a good opportunity to discuss the concept of "mapping" our starting materials and reactants. This concept, which is also sometimes called "overlaying", allows us to plot out which atoms from the starting materials appear in the final product. This can help guide us to the final product by giving hints as to where certain atoms should be located in the end. For this final product, the mapping is actually relatively simple.

We can place the starting materials and reactants next to each other, do a small rotation and label the atoms which might be significant. Once we have done this, the route to the final product becomes a little more transparent. As demonstrated above, we need to form a bond between atoms 1 and 2 and atoms 4 and 5 to create our fused ring. Because our map shows that we lose an oxygen atom (loss of "O"), but gain a double bond (loss of "2 H"), it stands to reason that we also must perform a dehydration somewhere (Loss of "2H" + Loss of "1O" at the same location = Loss of 1 H_2O). Atom 3 is labeled as a marker to remind us where this carbonyl comes from and that it will remain unchanged.

Now that we have an idea of which bonds need to be formed, we can start our mechanism rationalization by using the EASE method on the starting material and reactants.

In step 1, we see there are several possible electrophiles, but no real nucleophile yet. In step 2, we see there is a strong base in KOH and an alpha carbon which can be deprotonated on the ring. There are no steric issues, so we deprotonate and reassess.

We now have an enolate ion on our ring, and can start the EASE method again. In step 1, we see we have an enolate at atom 1 (refer back to the map) and a good electrophile on the alpha-beta unsaturated ketone (atom 2). In step 2 & 3, we see there are no acids/bases or steric issues, so we can move right to step 4, where the nucleophile attacks the atom 2 on the ketone as shown below.

This is the first connection toward creating the fused ring final product. If you refer back to our map from the previous page, we see that atoms 4 and 5 must now be connected to give us the final product. However, our new enolate does not have a negative charge at atom 4. How do we do this? The answer is found in resonance and base-catalyzed tautomers.

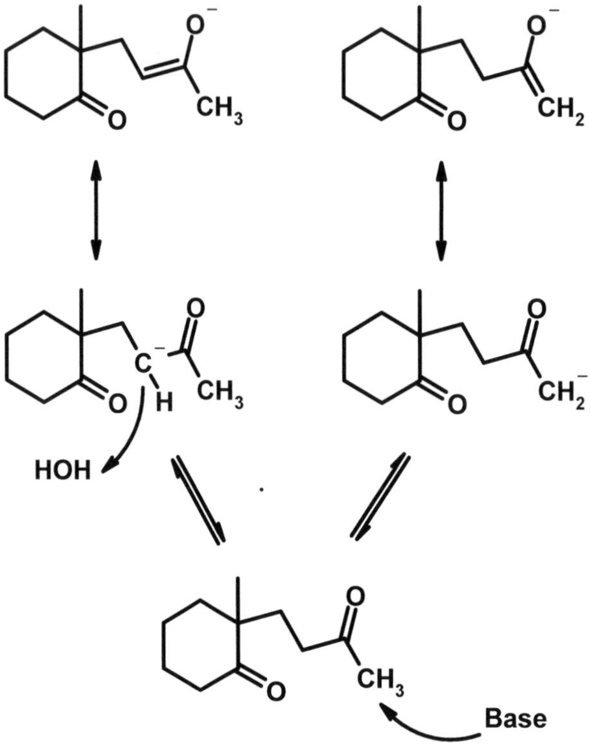

We can move electrons using resonance and atoms through a base-catalyzed tautomerization to show that a negative charge can be placed on atom 4, as shown in the lower structure in the diagram above. We are now very close to the final product. Again using the EASE method, we see that our nucleophile is the enolate on atom 4 and our electrophile is the carbonyl carbon (atom 5). In step 2 & 3, we see there are no acids/bases and sterics are actually working in our favor as atoms 4 and 5 are geographically close to each other, which can only help with the reaction. In step 4, we have the nucleophile of atom 4 attack the carbonyl of atom 5 and form our second crucial bond of the final product.

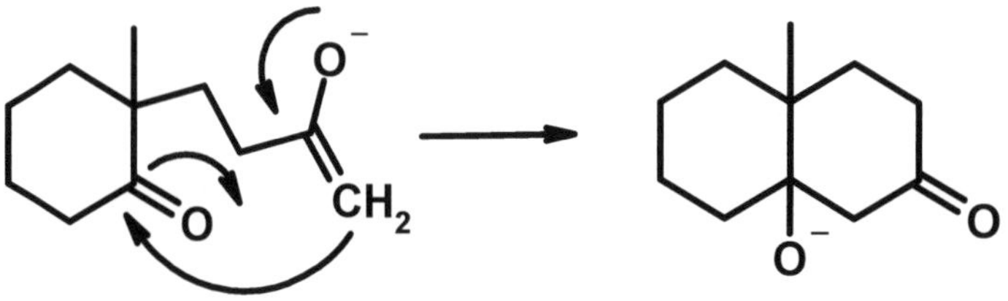

The final step of the mechanism is relatively easy and involves dehydration to form the new alpha-beta unsaturated fused ring system.

Overall, your mechanism "ring" will look like this:

FAQ: In the first step of the reaction, we deprotonated a carbon to give the enolate on the side of the methyl group. Why form the enolate on this side instead of deprotonating on the other side of the carbonyl?

Answer: This has to do with kinetic vs. thermodynamic enolates. Kinetic enolates are less substituted enolates and are formed faster than thermodynamic enolates. Thermodynamic enolates are more substituted and more stable but formed slower than their kinetic cousins. Reaction conditions dictate which enolate you will form, as shown below:

[Reaction 1: 2-methylcyclohexanone + LDA, THF, -78C → less substituted enolate] **Kinetic product**

[Reaction 2: 2-methylcyclohexanone + NaOCH$_2$CH$_3$, CH$_3$CH$_2$OH, Room temp → more substituted enolate] **Thermo product**

Kinetic enolates are formed using strong, non-nucleophilic (aka hindered) bases and polar aprotic solvents at low temperatures. Thermodynamic enolates are formed using strong bases and polar protic solvents at room temperature.

In our problem, we knew we needed the thermodynamic enolate to justify how we obtained the final product shown. However, you should remember that if you needed it, you could have also formed the other enolate, using different reaction conditions.

FAQ: Why did the enolate from atom 5 attack the carbonyl? Why didn't the other enolate attack?

Answer: If the other enolate had been formed and attacked the carbonyl, it would have created a four-membered fused ring. Not only is this a strained, unfavorable ring, it is also not what we knew was the product.

Correct Product

Strained Product

Easy Problem: What is the Product?

We start with a carboxylic acid and an alcohol in the presence of HCl. What happens? Our "E," the carboxylic acid, is a weak electrophile. The alcohol is a nucleophile. That is the easy part. HCl, which is definitely a strong acid, can donate a proton to the carbonyl, which can accept the proton on the oxygen. When thinking about step 3 we see there are no real steric issues.

E- Step 1: E+ is weak, ROH is nucleophilic
A- Step 2: HCl is def an acid, C=O can accept a proton
S- Step 3: No steric issues

Since the carboxylic acid is such a poor electrophile and we are in the presence of a strong acid, the first thing we should do is transfer the proton because that always happens first anyway, if a strong acid is present.

Continuing on, we now have a carbocation and an alcohol. This is pretty simple now. Our "E⁺" is the carbocation and the "Nu⁻" is the alcohol. There are no acids or bases left, as the HCl catalyst has been all used up (for the time being). Since we don't have any steric issues, the next course of action is to have the alcohol attack the carbocation. First, we have noted with the double-headed arrows that this is a reversible step. But second, we are getting very close to the final product now because it is clearer that all we have left is a couple of proton transfers and an ejection of water.

Strong E+   **Nu**

E- Step 1: E+ is carbocation, Nu- is ROH
A- Step 2: No acids or bases left
S- Step 3: No steric issues
E- Step 4: Nu- attacks E+

We perform a proton transfer from what was ethanol to one of the hydroxyl groups, turning it into water, which is a great leaving group. Remember, we still have chlorine floating around, which will act as a (weak) base because it is negatively charged. The chloride ion abstracts a proton, which causes the ejection of water, and gives us our final product. The reaction we've just shown is the Fischer esterification. It's a basic organic chemistry reaction that professors like to test you on. But this is a really good example of how you can figure out a problem by doing it in a step-wise and methodical fashion, using the EASE method. The Fischer esterification, at its essence, is a lot of proton transfers and one nucleophilic attack. Really, that's it.

And here's a hint for the future: When you see a geminal diol, which is two hydroxyls coming off same carbon atom, it will usually form into a carbonyl at some point.

Intermediate Problem: What is the Product?

[Thioester + HCl / CH₃OH reaction scheme]

In assessing this problem, we see that we have a thioester which is subjected to HCl and methanol. After we label all of the players, we see the following:

[Labeled scheme: carbonyl O = Base; C = E+; S = base; HCl = acid; CH₃OH = Nu]

For step 1, we know that the carbonyl is electrophilic and the methanol is nucleophilic. For step 2, the acid is obvious, but a corresponding base is not as clear. Both the oxygen and sulfur atoms have lone pairs, and could conceivably accept a proton, but they don't stand out as classic bases. Nonetheless, we should label them as such. For step 3, there are no steric issues, so we will move directly to step 4. As HCl is a strong acid, we need to move that proton first. While both the oxygen and sulfur could act as a base, the oxygen is the stronger base, as shown via resonance. Thus, the proton is accepted by the oxygen atom and we reassess using EASE.

[Mechanism arrow: O of thioester attacks H of HCl → protonated thioester with OH⁺ on carbon, S-ethyl intact]

The problem becomes much more obvious to us now that we have created the carbocation where the carbonyl used to be. In step 1, we see the carbocation is a great electrophile and methanol will do just fine as our nucleophile. [Quick note: Cl⁻ is still floating around, which is also a good nucleophile. In this problem, chloride ion will not attack the carbocation and give a final product, because it would place another good leaving group on the carbon atom, which would just be ejected at some point to give us another product.] Step 2 and 3 are not applicable as there are no acids/bases or steric issues, so we move directly to step 4, which is to have the methanol attack the carbocation.

Now the problem moves much more quickly. We regenerated HCl when the H from methanol was lost during attack, so we have a strong acid which is going to protonate something. While we have three candidates for protonation (the two oxygen atoms and the sulfur atom), the sulfur will win out. This is because protonation of the OH will not lead to any viable products, and protonation of the methanol will just reverse the attack of methanol which we just accomplished. That leaves sulfur as the atom to catch that proton.

Now that sulfur has a positive charge on it, we have a great leaving group. The final step of this problem is to re-generate a carbonyl (and our HCl catalyst) by ejecting the leaving group to give the final product shown below.

We went through this problem in painstaking detail, but you don't necessarily have to. We tried to show here that you can use the EASE method to start off, but if you feel comfortable moving forward after the first step or two without it, you should.

Real Chemistry You Have Never Seen Before:

Wow. This looks like a humdinger of a problem. This is our championship game. There are functional groups everywhere on this thing. This might be the hardest opponent you have faced yet. You can do this though. There is nothing here that you have not seen in another form somewhere else. Use the EASE method to break it down.

In step 1, we see there is a primary amine and a double bond. Both could be potential nucleophiles, but the amine is probably a more active nucleophile, so we will give it an "!" to show it might be stronger. The C=C reactivity is diminished because it is in conjugation with the C=O. As for electrophiles, there is an acid anhydride, a carbon attached to a leaving group and a conjugated ester. Of those, the anhydride and the leaving group carbon are the strongest electrophiles. In step 2, we find a Lewis base in the form of triethylamine (Et_3N), but no acids that stand out to us. Step 3 shows no horrible steric issues, except possibly for the bulky ether near the amine. We will keep it in mind, but it probably won't affect our reaction. Therefore, we can move to step 4 and have an electrophile attack a nucleophile. But which will attack which? It is a good bet that the amine is our strongest nucleophile and the fact that we are adding acetic anhydride is a good clue that it will attack that reagent, which is our electrophile. So let's move some electrons.

E- Step 1: Found them. C=C has no partner, COOEt is conjugated, amine and anhydride are good
A- Step 2: Not really, no acids only hindered Lewis base.
S- Step 3: No steric issues, we think
E- Step 4: Nu- attacks E+

The result of this reaction is that we have placed an acetate on our amine, which will now make it much less nucleophilic by converting it to the amide. This new group (abbreviated by "Ac") will now keep this amine from reacting when we don't want it to.

So, we've protected our amine as the amide, making it much less nucleophilic. Now, the second step of the synthesis is to subject our starting material to sodium azide. Again, let's label all of the reactants that we can. We still have a possible nucleophile in the alkene and somewhat of a weak electrophile in the form of the conjugated ester. We can ignore those both for now. Our strong electrophile is the carbon attached to the mesylate (-OMs), which is a great electrophile. We also have a strong nucleophile sitting there, in the form of NaN$_3$. In step 2 and 3, we see there are no acids/bases, and no huge steric issues. Therefore, we can move to step 4 and have the azide displace the mesylate in an SN2 fashion.

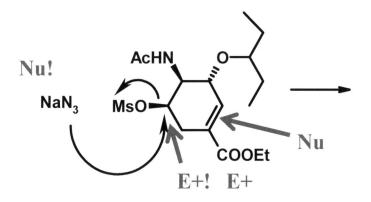

E- Step 1: C=C has no partner, COOEt is conjugated, azide and mesylate are good
A- Step 2: None
S- Step 3: No steric issues, we think
E- Step 4: Nu- attacks E+

Please notice, when the azide attacked at that carbon, the stereochemistry was inverted because it's an SN2-like attack. This provides our final product and the real-life chemistry that has occurred is part of the pharmaceutical pathway to create Tamiflu, a potent anti-viral drug. As you can see, you can rationalize your way through real-life chemistry that you have never seen before. Pretty cool.

2 steps away from Tamiflu

4th Quarter:
Summary, Study Tactics & Further Help

Your Organic Chemistry Toolbox:

To this point we've spent almost the entire book discussing how organic chemistry mechanisms work. However in most undergraduate classes, this is only half the game. The other half of the game is doing long multistep synthesis problems. In the average undergraduate class, these problems generally consist of 4 to 6 synthetic transformations, for which you don't need to know the mechanism.

There are two parts to mastering these multistep synthesis problems: performing a good retrosynthetic analysis and having a solid toolbox of reactions. In this section, we'll talk about that toolbox.

As we show on the next page, there are really only three types of organic chemistry reactions that should be in your toolbox. These are the reactions that form carbon-carbon bonds, reactions that form "carbon-X" bonds, and reactions that convert one functional group into another.

The simple way to think about these three types of reactions is:

1) Carbon-Carbon Bond Forming Reactions: Put two carbon atoms together.
2) Carbon-X Bond Forming Reactions: Put a new functional group on carbon, where "X" is not necessarily a halogen.
3) Functional Group Interconversion: Change one functional group into another.

Generally when you start a problem, the professor will tell you what type of starting material you can use. Almost always this is going to be fewer carbons than your final product contains. Therefore, somewhere in the transformations you will have to execute a carbon–carbon bond forming reaction. And then to make sure you know the concepts, they always throw the other two types of reactions in there.

So without further delay, we present the AceOrganicChem.com toolbox of organic chemistry reactions. As stated above, it is organized into three different types of reactions. And just like you would place the tool you use the most top of your toolbox, we have placed the reactions we think you will use most of the top chart. We think these are reactions that every organic chemistry student should know and be able to use quickly.

AceOrganicChem's Organic Chemistry Toolbox

Carbon-Carbon Bond Forming Reactions	Carbon-X Bond Forming Reactions	Functional Group Interconversion
Grignard addition	**Halogen addition to an alkane**	**Oxidation of alcohols**
Alkyne addition	**All Electrophilic Aromatic Substitutions**	**Reduction of carbonyls**
Enolate Addition	**Grignard Formation** $CH_3Cl \xrightarrow{Mg} CH_3MgCl$	**Hydrogenation:**
Wittig Reaction *	**Epoxidation**	**Hydroxyl to chloride**
Aryl Ring Additions, i.e. Friedel Craft alkylations	**Hydration of Double Bond**	**Esterification**
Diels-Alder Reaction	**Ozonolysis**	**Carboxylic acid to acid chloride**
Cyanide Addition	**Dihydroxylation**	**Nitro group to an amine**

Let's Go Retro: Retrosynthesis

Ah, remember the 70's? Discos, a nice large 'Fro, and socks. Retrosynthesis is kinda like the 70's, except it actually has a purpose. The purpose of retrosynthesis is to give us a viable pathway to get from a small molecule to a large one. The theory behind it is that it is much easier to breakdown a large molecule to see which small molecule it could come from than the other way around. There are really two actions that can be taken in a retrosynthesis, as shown here:

Synthons:

CAUTION: Synthons are not necessarily real

Functional Group Interconversion

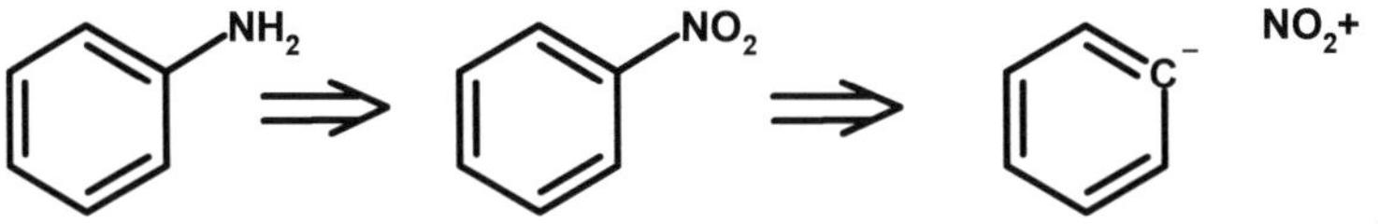

Synthons are the fragments that come from breaking a bond. The important point with synthons is to recognize that they are not necessarily real. They just represent the fragments which might be possible. For example:

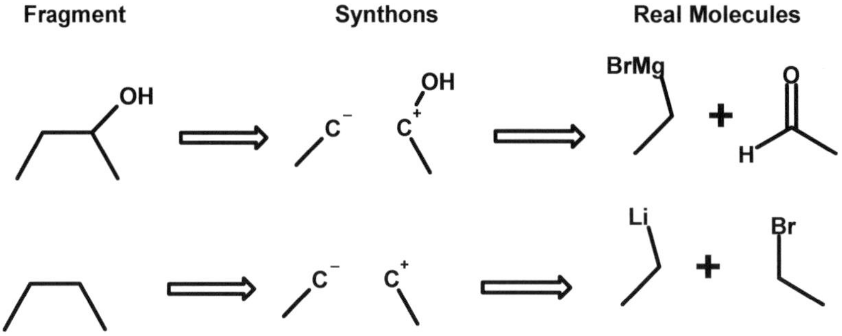

Fragment   Synthons   Real Molecules

Here, we are cleaving strategic bonds. We assign a hypothetical "+" or "-" charge to each end of the broken bond to help us decide which real molecules our synthons should come from.

Once your major carbon-carbon bonds have been formed, it is time to change the functional groups in the groups we need for our final product. Functional group interconversion (FGI) in retrosynthesis is the reverse reaction of all the FGI that you have in your toolbox. Unfortunately we don't have the time or space to discuss this subject at great length, however suffice it to say this is a very important part of organic chemistry.

While entire books have been written just about retrosynthesis, here are some simple tips to help you complete your synthesis problems more effectively:

- ✓ DO: Work backwards to an "acceptable" starting material
- ✓ DO: Count your carbons so they don't get lost or forgotten
- ✓ DO: Locate functional groups, they will be the first & best place to look for disconnections
- ✓ DO: Place most reactive functional groups on product last
- ✓ DO: Examine more than one pathway (time permitting)
- ✓ DO: If you run into trouble, start interconverting functional groups looking for ideas.

- ✗ DON'T: Ignore other functional groups on the molecule.
- ✗ DON'T: Make up chemistry that "looks like it should work."
- ✗ DON'T: Use reactions that will give mixtures of product. (May lose points on exam)

Retro FAQ's:

FAQ: Where should I choose to disconnect?

Answer: More often than not, disconnections take place immediately adjacent to functional groups because they have electronegativity differences from carbon and therefore can do chemistry.

FAQ: How do I recognize a good disconnection?

Answer: A good disconnection will make the molecule visibly simpler.

FAQ: How do I decide which synthon carries the charge?

Answer: One trick is to use resonance. If you draw a synthon and the resonance structure looks similar to a real reactivity intermediate, you are most likely on the right track.

FAQ: Was disco ever cool?

Answer: No, they just didn't know any better back then.

Spiderwebs:

This is the demonstration of true organic chemistry mastery. Well webslingers, we are almost to the end. We have found that this to be a powerful study aide for students at almost any time in the semester. All it entails is creating a "link chart" between functional groups. Start with a simple alkane and think of all of the reactions you know on alkanes. Then take alkyl halides and figure out all of the reactions you know for those. Pretty soon you will see that you are finding multiple methods to get to the same functional group and multiple reactions you can do starting with a certain functional group. Your lines will start to cross a lot and you will see that all of the functional groups are interrelated. This is the essence of functional group interconversion (FGI), and is a very useful tool to master. Below, we have placed a sample one for you to get started on. Be sure to place conditions for each reaction somewhere in the web while you are doing it. Your web should be different from ours based on the different reactions which are important to your class. The wonderful thing about this exercise it is will help you study and simplify synthesis problems when the time comes.

Truth be told, this is a great exercise to do with some friends, a white board and some cold beverages one night.

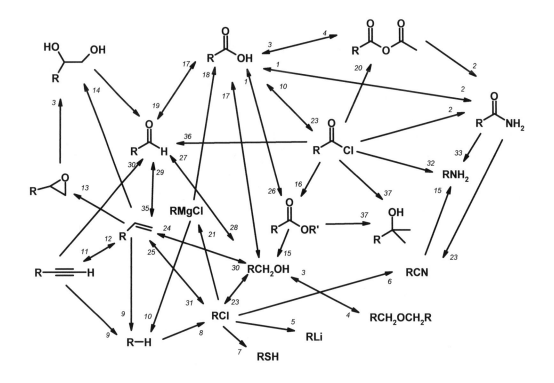

1. NaOH
2. NH₃, H₂O
3. H⁺, H₂O
4. Heat, loss of H₂O
5. 2 Li
6. NaCN
7. H₂S
8. Cl₂, hv
9. H₂, Pd-C
10. H₂O
11. Br₂, then NaNH₂
12. H₂, "Lindlar's" cat
13. mcpba
14. OsO₄
15. LiAlH₄
16. ROH
17. K₂Cr₂O₇/H₂SO₄
18. CO₂
19. DiBAl-H
20. CH₃COONa
21. Mg, Et₂O
22. NaOH, then LiAlH₄
23. SOCl₂
24. H₃PO₄, heat
25. NaOMe, MeOH
26. ROH, H⁺
27. PCC or PDC
28. NaBH₄ or NaCNBH₄
29. O₃, DMS
30. BH₃,THF then H₂O₂
31. HCl
32. NaN₃, then NaOH/H₂O/heat
33. Br₂, 2 NaOH
34. NH₃, NaCNBH₄
35. Ph₃P=CH₂
36. LiAlH(O-tBu)₃
37. excess CH₃MgBr

Overtime:

What Not To Do With EASE

Where the method will be less effective:

Through the course of this book, we have tried to show you a systematic way to look at many organic chemistry reactions to figure out how they are working and what your reaction product would be. The system works because it forces you to examine why the reactions would proceed a certain way and tries to keep you from just giving a product that looks like it should be right. When the system works, it is a beautiful thing. But now that you are an organic chemistry all-star, we need to have a tough conversation about when it doesn't make sense to use the EASE method. There are several classes of reactions that do not work easily with the EASE method. It's not that they won't work with the EASE method (it is still **_always_** about electron flow), as much as it is that it is easier just to learn what the reactant is and what it does rather than why it does it in some cases.

Oxidation and Reduction:

The first type of reactions that you should not use the EASE method for are most reduction and oxidation (RedOx) reactions. Again, we want to stress that the method could work for these reactions, but it is just easier to memorize a couple of reagents and apply them, rather than slogging through the whole EASE method.

As you learned in general chemistry, losing electrons is oxidation and gaining electrons is reduction. The mnemonic for this is "LEO the Lion goes GER"...LEO = Losing Electrons is Oxidiation, GER = Gaining Electrons is Reduction. There are a limited number of RedOx reagents and reactions that you will have to familiar with, so it makes sense to just learn them right now. Here are some:

Reaction	SM	Product	Reagents	Note
Oxidation	1° Alcohol	carb acid	CrO_3/H_2SO_4	Harsh oxidation
Oxidation	1° Alcohol	aldehyde	PCC or PDC	Soft oxidation
Oxidation	Alkene	diol	OsO_4 tBuOOH	Dihydroxylation to cis diol
Oxidation	Alkene	ketone	O_3, Zn	Ozonolysis
Reduction	Ketone	2° alcohol	$LiAlH_4$	Harsh reduction
Reduction	Aldehyde	1° alcohol	$NaCNBH_4$	Soft reduction
Reduction	Alkyne	E-alkene	Na, NH_3	Gives t-alkene
Reduction	Alkyne	Z-alkene	Lindlar	Give cis alkene
Reduction	Alkene	alkane	H_2 / Pd-C	Full reduction

"SM" = Starting Material. We have tried not to overwhelm you with these reactions, as there are actually many more than on this list which will accomplish a range of RedOx reactions. As stated above, the EASE method will work just fine with many of these RedOx reactions. We think that it is easier for you to just know these reactions, as opposed to figuring your way through each one. Take the reaction below:

If you were going to do the full EASE method for this problem, you would see for step 1 you have a carbonyl as your electrophile and $NaBH_4$ as your nucleophile. Remember that $NaBH_4$ is sodium borohydride, and that hydride is H⁻ and can act as a nucleophile. In step 2, we see that hydride can be a base, but we really don't have an acid, as it is not strong enough to deprotonate at the alpha carbon of the carbonyl. There are no steric issues, so we can move to step 4 and have hydride attack the carbonyl to give us the intermediate shown. This intermediate will then find a proton from the water floating around to give the final product. The overall reaction is the reduction of the aldehyde to an alcohol.

OR.....you could just remember that NaBH$_4$ reduces aldehydes to alcohols and do it all in one step without using the EASE method. The upside to *not* using EASE here is you gain a little of time on your exam. The downside to not using EASE here is you might miss a nuance in the reaction which would give you a different product. It is up to you when you think it is best to use the EASE method. We just want you to be aware that there are circumstances where it might benefit you not to use it.

Radical Reactions:

Radicals are electron deficient species. While technically one could argue that this places them in the electrophile category, they don't behave like classical electrophiles, which makes them poor candidates for the EASE method. Since we are trying to be brief with this book, we will not go into radical chemistry here. Suffice it to say that you if learn the three main types of radical steps (initiation, propagation, and termination), and several of the major types of radical reactions (shown below), you will be just fine with this and not need to use the EASE method.

Organometallic Reagent Formation:

This is one more place where you don't need to rely on the EASE method, you should just know the reagents and what they form. There are four major types of organometallic reagents you should know, all shown in the figure below. As the metal insertion of these reactions in some cases behaves a lot like a radical species, it is just easier to know that the metal insertion occurs and not worry about how or why. [Note: "X" is a halogen, almost always Cl, Br, or I.]

$$\text{R-X} \xrightarrow{\text{Mg}} \text{R-MgX}$$

$$\text{R-X} \xrightarrow{\text{Li}} \text{R-Li}$$

$$\text{R-Li} \xrightarrow{\text{CuX}} \text{R-Cu-R}$$

$$\text{R-X} \xrightarrow{\text{Zn}} \text{R-ZnX}$$

So there you have it. The EASE method for organic chemistry is designed to force you to think about the basic tenets of organic chemistry when you look at a problem. The most basic elements of any organic chemistry problem never change: 1) Find out who has the electrons, 2) label any acids or bases around and move the proton if you can, 3) identify any steric issues, and 4) let the electrons flow. Then rinse and repeat, if needed. The beauty of the method is that you learn a step-wise methodology which can be applied in almost any organic chemistry situation.

Hopefully you found this helpful. We take feedback (questions, comments, complements and/or hate mail) at **admin@aceorganicchem.com**. Thanks a lot for purchasing this book, good luck with your studies and happy reacting.

Appendix 1: Answers To Supplemental Exercises

Exercise 1.1 & 1.2

If there is no arrow, that means it is not a nucleophile or electrophile.

Exercise 2.1

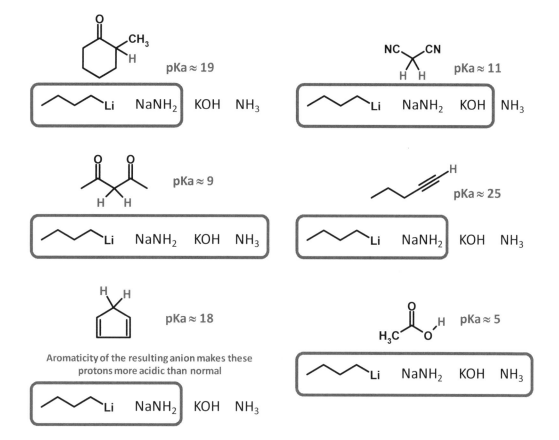

Exercise 2.2

Explanation of Exercise 2.2

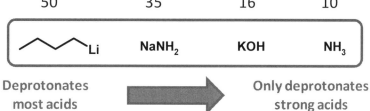

Exercise 2.3

KOH — Strong Base

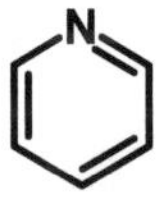

HCl — Strong Acid

pyridine — Weak Base

NaHCO$_3$ — Weak Base

HNO$_3$ — Strong Acid

H$_3$O$^+$ — Strong Acid

HCN — Weak Acid

CH$_3$CH$_2$ONa — Strong Base

CH$_3$CH$_2$CH$_2$CH$_2$Li — STRONG Base

Exercise 3.1

Least   **Kinda**   **Most**

Bulkiness decreases nucleophilicity. In fact, t-butyl oxide is essentially non-nucleophilic.

Exercise 3.2

The isopropyl group has the least amount of hindrance of the four of these compounds

Exercise 4.1

This is an example of an acid-catalyzed dehydration

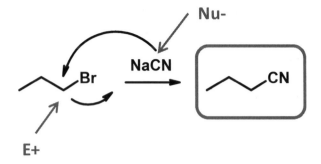

This is an example of an S_N2 reaction.

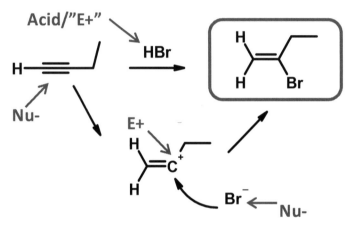

Addition to an alkyne.

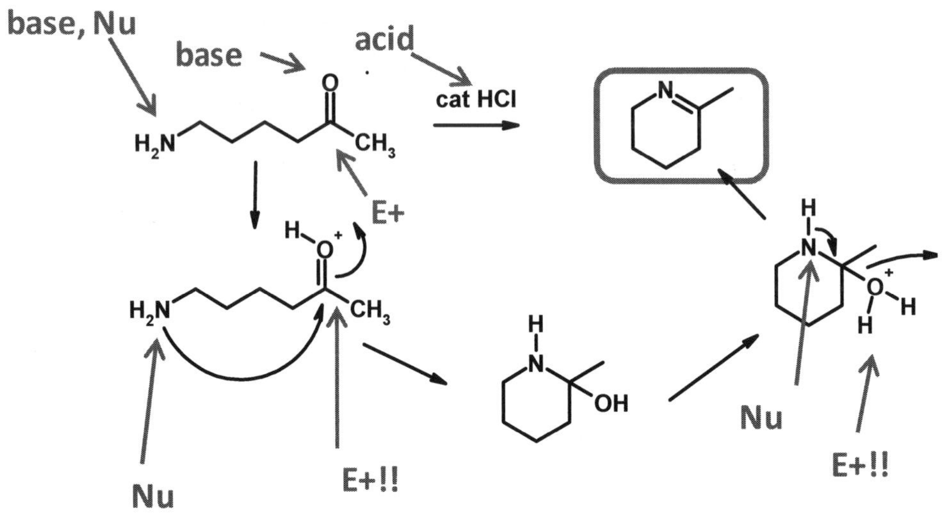

base, Nu
base
acid
cat HCl
E+
Nu
E+!!
Nu
E+!!

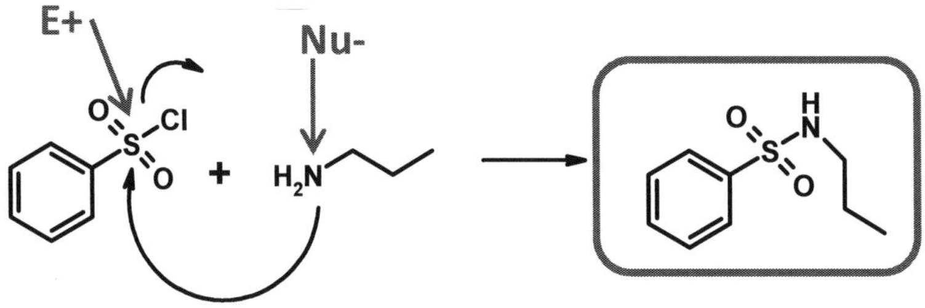

E+
Nu-

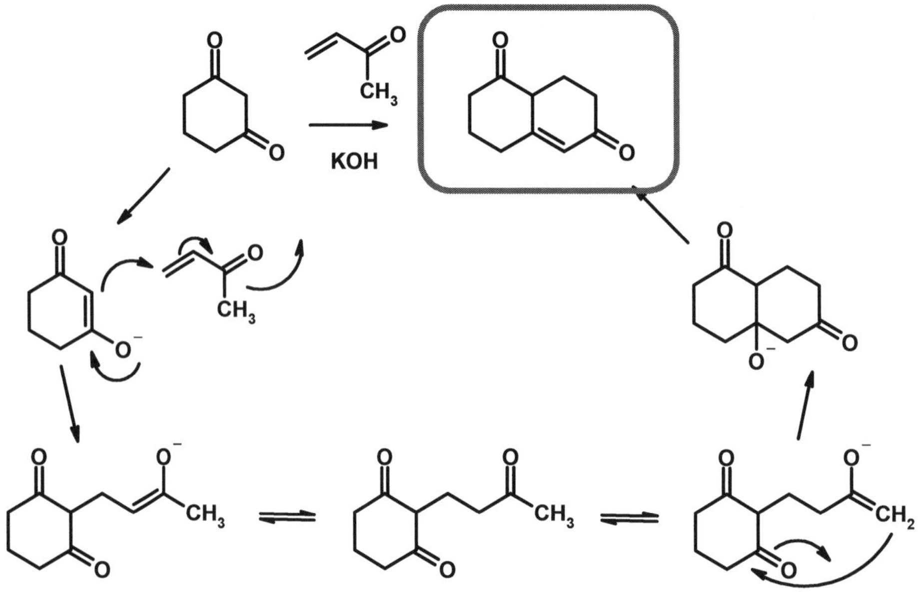

KOH

For Your Notes:

For Your Notes:

For Your Notes:

For Your Notes:

For Your Notes:

For Your Notes:

For Your Notes:

For Your Notes:

For Your Notes:

For Your Notes:

For Your Notes:

For Your Notes:

For Your Notes:

Made in the USA
Middletown, DE
01 April 2015